FLIPPHONIA

Combatting Mobile Social-
Media Addiction in America

JASON O'NEIL

authorHOUSE®

AuthorHouse™
1663 Liberty Drive
Bloomington, IN 47403
www.authorhouse.com
Phone: 833-262-8899

Published by AuthorHouse 12/14/2022

ISBN: 978-1-6655-7807-3 (sc)
ISBN: 978-1-6655-7806-6 (e)

Library of Congress Control Number: 2022923126

CONTENTS

BOOKS BY THE AUTHOR

Bald Eagle Vision
The Red Box
Turbopod
Turbospace
Sinecure
Cyberclipper
When Baldie Cries
New Age Ark
A Necessary Coup
Micronations
Hypersonica
Mission Embryo
DroneViper
Bald Eagle Vision 2
DragonChip
Cellphonica
Solarmania
Escape from La-La Land
Mirachip
Tough Sale
Padlockers

CAST OF CHARACTERS

Matt Flynn
Company Founder, billionaire entrepreneur
Age 72
Married to Heather
Similar to: Burt Lancaster, Actor

Heather Flynn
Aeronautical Engineer
Age 70
Project Manager
Similar to: Grace Kelly of Monaco

Murray Flynn
Son of Flynns
Project Manager
Age 40
Similar to: Young Harrison Ford, Actor

Maggie Flynn
Married to Murray
Age 38
Social-Media Guru
Similar to: Amy Adams, Actress

Cole Langford
Cyber Guru, Co-founder of Padlockers
Married to Dawn
Age 42
Similar to: Young Neil Armstrong, Astronaut

Dawn Langford
Cyber Guru, Co-founder of Padlockers
Married to Cole
Age: 39
Similar to: Michaela Schriffrin, Skier

Prince Yousif Latif
Prince of Dubai
Business partner of Flynns, billionaire
Age 65
Similar to: Omar Sharif, Actor

Chin Chin Po
Intelligence Analyst/Spy
Age 50
Partner of Andrew
Similar to: Zizi Zhang, Actress

Andrew Ho
Famous Architect/Spy
Age 52
Partner of Chin Chin
Similar to: Bruce Lee, Actor

Alexander Murray
President of Voice Synthesis, Inc. (VSI)
Age 56
Similar to: Rory McElroy, Golfer

Tinshin Wei
MIT Professor, Diplomat
Age 48
Similar to: Li Bing Bing, Supermodel

Major General Thomas Taylor (USA)
Asst. Secretary of Defense
Age: 64
Similar to: Gen. Norman Schwartzkopf

1

MIAMI

Billionaire entrepreneur Matt Flynn and his wife Heather and their son, Murray, and his wife, Maggie invited six people to their Key Biscayne mansion to talk about a subject on their mind: the destruction of America by social media. The guests included a Chinese couple from Hong Kong, Chin Chin Po and her partner, Andrew Ho. The Langfords, Cole and Dawn, had just arrived from Washington, D.C. Joining the group was Alexander Murray, President of several non-profit entities. And last to join was Tinshin Wei, a tenured professor at MIT who is an expert on the Chinese Communist Party CCP). With adult libations in hand, the group sat in a ring in a gazebo next to the swimming pool.

"Thank you for coming on such short notice," said Matt as he opened a two-day strategy session. Heather and I have a true concern for the direction of our country....away from a free-enterprise economy and democratic political system created by our Constitution. In a moment I'll show some revealing, indeed shocking, statistics about this crisis. But suffice it to say that across this country our citizens of all ages are consumed by the digital world taking them away from the real world. This crisis is the direct result of the cell phone over the last thirty years. And today it is a national addiction to mobile smart phones. We've amassed a sizeable reference library, done a lot of reading and even had experts as guests to confirm our findings. Heather has just given you a list of some, and I stress only SOME, of the impacts of this cultural revolution. Please take a moment to read the list and feel free to add some, provided you put your smart phone in airplane mode.

(Ha-Ha)

- Pre-occupation at the expense of the work ethic leading to "Quiet Quitting"
- Belief in a fantasy world of La-La Land to avoid lifetime decisions
- Cryptic text messages replace English grammar and cursive writing
- Consensus that physical possessions are unnecessary
- On-Line relationships are accepted as substitutes for the real- world ones
- More meaningful communication via digital text even though people are sitting next to each other
- Lowering the standard of living for everyone by the loss of friends, family, work and local causes
- Lessens the need for parenting and the youth obeys social media not their parents
- Removes meaningful role models to encourage and enable the escape from La-La Land
- Neglect of pets; walking Fido is too time-consuming

- Enjoyment of spending time doing non-productive tasks. "I hide in my video game while someone else, particularly the Federal Government, takes care of my needs."
- Trapped by newer, more capable devices, cheaper storage and creative applications such as video streaming and wagering.

"Wait a minute, Matt, your talking about a global disease for which to date there is no cure," said Alexander Murray, a successful executive and long-time friend of the Flynns.

"My friend, you're absolutely right. And like COVID-19, this disease comes from and is spread by China as an element of the country's long-term plan to dumb down our society in advance of a totalitarian takeover."

The two friends from Hong Kong nodded their agreement with the statement.

"But more on that later. Now Heather is handing out a data list showing just how bad the situation really is. Please read each point and let the impact sink in. And I warn you. This is not a pretty picture which is emerging, just like our Asian adversary wants it to be. Addiction starts at age 10 where a social media device is justified by parents as a safety tool for travel to and from school. Look at the results:

- 85% of Americans own a smart phone
- 58% of Americans say their cell phone is "my most valuable asset."

A gasp went around the gazebo.

- 47% say: "I'm addicted to my smart phone." (It's growing by 10% a year.)
- 53% would save the cell phone first in a house fire
- 53% have never gone longer than 24-hours without a cell phone
- 61% have texted someone else in the same room
- 62% sleep with the phone
- 64% use the phone while on the toilet
- 35% use the phone while driving
- 43% use the phone while on a date
- 46% spend more time on the phone than with their significant other
- 49% feel stressed if the smart phone is lost more than 30-minutes
- 75% routinely access multiple application sites daily. The most popular are:

Facebook (71% of the market)
Snapchat
Twitter
Instagram
Nextdoor
Tiktok (Chinese-owned)
Pinterest
Linked-In
And dozens more.

- 95% of college students are "hooked" on mobile social media, often to promote "ultra-liberal" causes
- 90% feel compelled to answer a "ping"
- 25% go into debt for the latest smart phone
- 33% would give up a pet to keep a smart phone

As Matt petted the family's Golden Retriever, Lily, he said, "It's that last one that really got me!"

Everyone nodded their understanding

Then Heather said, "Wait, my friends, it gets even worse."

"Worse! How is that possible?" asked Andrew.

"On average, Americans check their phone 340 times a day or once every 4-minutes. And this statistic does not include time on smart media at school or while doing homework."

"Income is not a major differentiator. In many cities, the youth are provided digital devices, usually laptops, for distance learning. Given the drastic drop in standardized test scores for both Math and English, it's clear that social media is primarily being used for games and entertainment."

"Heather, you're right! Both Baltimore and the District are classic examples of the depth of the problem," exclaimed Dawn.

"Wait, my friends, it gets EVEN WORSE! Did you know that teens spend more time on social media than they spend sleeping or with their parents. National polls confirm their desire to text rather than have a one-on-one conversation where they would have to "look the other person in the eye."

"So, my friends," interrupted Matt, "this does not bode well for generating national leaders who can communication effectively, much less sign their name in cursive!"

"So, it's 11:00...time for a swim and then lunch. We'll get back on the subject this afternoon."

Broad smiles and nodding heads circled the serene garden structure.

2

MIAMI II

At 3:00 PM the group reconvened in the library for a round table discussion of potential solutions. Each person brought a unique and valuable set of skills to the discussion. The host kicked the session off by saying, "Guys, as I mentioned, Heather and I have been studying this topic for several months. But we do not want to unduly influence the discussion. We can't return to the 1950's when life in America was perfect. But we're dealing with a crisis, and something has to be done soon. We're too close to the problem so we've requested that Murray lead the discussions. He's very good at asking dumb questions!"

(Ha-Ha)

"At 4:30 we'll break to dress for dinner and reconvene here for Happy Hour. Dinner will be informal at the Key Biscayne Yacht Club. Your input into the discussion will be truly appreciated. Maggie has consented to take notes. Murray, you've got the microphone, my son."

"Thanks, dad. It's an honor to participate in this discussion with so many special guests."

(Murray nodded at the guests and waved his arm in a semi-circle to recognize the talent seated in the book-lined room.)

"Maggie and I would like to initiate this round table with a few words about something which is tragically disappearing from our country: WORK ETHIC.

Thanks to my father and mother's ingenuity and business acumen, Flynn Enterprises has grown to 26 companies and just over 100,000 employees around the world. Maggie and I have been privileged to run a couple of them, fortunately at a profit. Our factories in the Philippines and Singapore and our design studio in Israel are incredibly productive. The workforces are loyal, eager to learn and extremely conscious of performing in a superior manner.

Tragically, I can't say the same thing for our facilities here. Our work conditions are the best on the planet. Our employee packages are rated #1. We've eliminated the need for unions which can suck the life out an enterprise. However, dispite all this, the work ethic is very bad and getting worse!

Maggie distributed a list to everyone.

"Please take a moment to review the list before we open the discussion:

- Lack of interview discipline; won't even show up for on-line interviews!
- Excessive tardiness with asinine excuses
- High injury rate based upon only minor distractions
- Constant demands for additional holidays, often for political reasons

- Constant demands for training in disciplines unrelated to the person's work responsibilities
- Constant demands for expanded break and lunch periods

The bulk of the employees are focused on maximizing benefits prior to an early retirement.

And a real sore point for management is that these longer break periods are for accessing mobile social media (S-M), and people are aggravated if their S-M sessions are interrupted for any reason...like going back to work. This is in part due to the fact that if an assembly line worker is caught looking at S-M, he or she is immediately fired."

"Finally, let me say that it's not a salary issue. It's a bitterness that their job is interfering with their social life style! And when we interview an employee as part of the person's annual performance assessment, it's clear that they know no other approach to life. One question we ask is: "What is the latest book you've read?" The person is stunned to silence. The senior managers simply shake their heads in disbelief."

Cole Langford raised his hand and said, "Murray, if you think it's bad on the factory floor, you would be revolted by the bureaucratic environment in Washington. Dawn and I could entertain you for hours with stories based upon our three-year project to close departments as the infamous Padlockers.

"Cole, I'm sure you could. But you and Dawn are national-known cyber gurus.

"With so many in Googleland all day long, it's a paradise for hackers. They create applications which look legitimate, but once you open it, your data is gone, usually forever. Smart phones are not smart enough to detect and neutralize the hackers. The losses are measured in hundreds of billions of dollars every year. Many people are financially ruined. And it's no secret these devices, primarily made in China, have covert backdoors to capture the data for future use to control our nation."

"Dawn, what's foremost in your mind about this digital pandemic?"

"Murray, Cole's right about the bureaucratic work ethic. I have many stories where this is particularly the case in the Defense Department where both "Govies" and their contractors hide behind high security clearances to conceal reduced work ethics. That's way we were able to reduce the headcount of several classified agencies by 50%. In addition, as you know, I've written a couple books about cyberbullying. It's a real problem only getting worse. Matt has asked me to say a few words about the topic at breakfast tomorrow. It highlights so many of the bad aspects of our rampant S-M."

"Thanks, Dawn. I'm really looking forward to your remarks."

"Alex, another aspect of S-M is Big Tech's ability to control public opinion, even sway national elections. Would you please comment on your experience and provide timely observations?"

"Sure, Murray, I'd be glad to. First, let me say that S-M creates an instant digital mob to support a candidate or firestorm issue like abortion. In our best American tradition, please challenge the following facts and observations:

- Big Tech furthers totalitarian censorship and can apply Big Data to suppress civil liberties
- With enough money, incumbency is virtually guaranteed and/or veto power paralyzes government
- The politician and/or Political Party which can't leverage S-M will always lose
- Major S-M platforms can instantly "force" the citizenry to have an opinion. "Alexa, what's my opinion on reducing the National Debt?"
- Big Tech always moves faster than federal or state regulations because the politicians use the media and don't want change
- Advanced surveillance can bust a company or person in an instant
- Artificial Intelligence (AI) creates "Deepfake" videos to sway opinion (Soon we will need Personal AI to discover the truth.)

"So, what we see," continued Alex, as he counted on his fingers, "are three key conditions":

- Big Technology can erode Democracy
- S-M proves the impossibility of developing perfectly secure software, networks and devices. So, cybersecurity becomes a central political issue and

– Technology providers, such as China, are able to mold opinion by dumb downing the electorate to submit to authoritarian control."

Alex ended his comments with a question to highlight his observations: "Alexa, should I vote, and who should I vote for Governor of Florida?"

The "Chosen Dozen" met in the library at 5:30 for flutes of France's Finest. The mood was very upbeat in spite of the magnitude of the problem. It was clear that S-M impacts everyone. "After all, 95% of the population owns a smart phone."

A six-course dinner in the Skipjack Room of the yacht club was an epicurean delight. Several toasts were made to the host for his generosity and ever-forward entrepreneurial endeavors. During dessert, Matt introduced his dear friend Tinshin Wei, the famous professor from the Massachusetts Institute of Technology.

The petite woman went to the podium, adjusted the microphone, smiled and said, "Thank you for this opportunity. I have consulted with both Chin Chin and Andrew about my findings and conclusions. My recent research exposes a lot about the use of S-M in China today.

And I do this because it's a key tool for their Five Year Plan to takeover of our country."

Professor Wei laid out her findings in a twelve-minute speech which highlighted the following points:

- Millions of cameras track tens of millions of people
- If a camera records a violation, however small, "Social Credit" rights and privileges are taken away and personal movement within the country is severely limited
- The goal of the Social Credit System (SCS) is to control everyone
- Watching Fake News from overseas via S-M is particularly bad
- The Social Credit System breeds a Bribery Culture for citizens to pay someone to avoid the loss of social mobility

So, in summary, the one a one-half billion citizens are slaves to the government cameras of the SCS. They are less than pawns on the chessboard of life.

Once a person offends, he or she (and their family!!) are blackballed. It is so bad that a person is blackballed when seen with a known blackballed person. It's a new form of Russian Communism and Stalin's Gulag System or weeding out any possible political dissent."

In the limousines on the way back to the Flynn compound, the discussion was about an urgency to limit the rise of S-M in America so SCS never becomes a tool of the Chinese in America.

The next morning a buffet breakfast was served in the conservatory. The guests remarked how refreshing it was not to have to answer a smart phone. No "pings" were allowed amidst the incredible variety of plants. At 9:00 Matt convened the morning session poolside. His instructions were quite simple: "We need a spirited discussion about potential solutions. Then early this afternoon we'll craft the major principals of a plan or charter for a new company which I have in mind."

(Hear, Hear and Bravo!)

Maggie Flynn kicked off the discussion by saying, "In my opinion, the detox process has to start at home. This is a parental challenge while they lead by example by having times of the day free from social media for personal think time and family activities. A ten-year-old can have a digital safety device for the trip to and from school. But that's it. If the device is more capable, the youngster will be willing, indeed pressured by peers, to use it school and beyond."

"I agree, "said Chin Chin." This is a family responsibility to give the youngster half-a-chance to start life without a digital addiction. In Hong Kong, some families take this responsibility so seriously that the devices are put in a lock-box certain times of the day. This is particularly important at dinner time to avoid the propaganda blasts from Beijing."

Several of the guests nodded their understanding of the problem.

"Well, I'm particularly alarmed at how Washington has used Big Tech's S-M platforms to spew disinformation, downright lies, to remain in power. It's not only unconstitutional but anti-American as well," added Dawn.

Cole nodded his agreement with the statement and added, "Soon we will need artificial intelligence applications on our devices to detect and counter this stuff."

Tinshin then added, "Maybe we need these lock-boxes at school. And we should be able to penalize the teacher's unions who promote the use of these S-M devices in order to have a bullhorn on behalf of the Democrat Party. Any teacher who tries to quiet the bullhorn is called a "racist." The situation at school is also ripe for students to bully classmates. Under these circumstances, America loses!"

Alexander then asked, "Why can't there be a national incentivized device confiscation program? People can be rewarded for turning in a smart device for a dumb one. The collected device can either be recycled or sold at a discount to Third World countries with all the proper warnings about their use. Proceeds from the sale would offset the cost of the program."

"Alex, that's a creative concept we can build upon later," said Matt.

Heather then added, "I believe we're right about the role of parenting. But let's face it, both grade and high school are the primary culprits in this whole affair. Can't we set up a program where teachers are rewarded for weaning their students off of S-M? This could also help neutralize the impact of their unions."

Nodding heads went around the group.

Architect Ho added: "You know each home today has Alexa. It's an extension of Amazonia. Couldn't we require Amazon to imbed anti-S-M messages in Alexa responses as a public service such as: "This music is nice, but now it's time to do your homework."

Everyone chuckled but appreciated the power of the idea.

Tinshin then jumped back into the discussion. "You know every opportunity to access a S-M device is an opportunity for the Chinese to dumb down our citizens. Fake news is pure propaganda. Therefore, we must find creative ways to limit the importation of these devices into our country. It's no different than illegal drugs made in China and smuggled here via Mexico."

Murray then added: "Tinshin, you're absolutely right! Maybe the answer is taxation and other penalties to slow the importation of restricted classes of S-M devices. And we must conduct periodic inspections to ensure the imported devices have no covert backdoors gathering data for the CCP."

Murray looked at Cole for a reaction.

"Well, we certainly have the technology to screen every device that comes from China. But the real solution would be to allow us to monitor the production process on the assembly line in China. But that won't happen, so the next best thing is to perform a 100% inspection at the port. This would slow down the pipeline so drastically that China might divert devices to other global customers. Such a solution, however, would be so distasteful to the public, I envision marches, riots and even wholesale robberies at the ports. Perhaps there is a clandestine means to disrupt the production. One might even consider paying our Taiwanese allies to not produce the required chips for the Mainland. We could also incentivize them by the production of Flip Phones."

"Wow, Cole, you have put some real thought into this problem. Somewhere thar's a pony in them thar hills!"

Everyone laughed.

At this point, Matt looked at his Patek Philippe watch and pronounced that it was time for lunch.

"My friends, this session has been as productive as I had imagined. We'll reconvene at 2:00 PM for a focused one-hour meeting to distill our thoughts down to a draft Action Plan, ideally implemented by the corporation I mentioned. At 4:30, please be my guest on my Wally118 boat as we welcome the sunset and return for dinner. I promise not to keep you up too late this evening. And I warn you. If you're caught using a smart phone on the boat or at dinner, you owe the group an apology.

Promptly at 2:00 PM, the group met in a covered terrace next to a putting green. Murray started the meeting by asking his father to describe the company he had in mind.

"Sure. Today, as your discussion has correctly highlighted, America is addicted to mobile smart devices made in China. And, today, China uses them to control its citizens. Heather and I interviewed doctors who compared these devices to slot machines which can change behavior like a dopamine high. And the S-M can easily create toxic relationships. Youth interact with their smart phones more than their own siblings! And there are hundreds of "scientific" studies which prove the value of multitasking while on the phone. These are blatant lies. It doesn't exist. It forces one to turn inward, and by age 18, the youth routinely turn inward extracting themselves from the world around them.

Therefore, we need a national effort to combat these digital handcuffs. A new corporation can help develop a new national communications infrastructure to combat this addiction. Heather is handing out a graphic which shows the basic organization of this new entity which we have trademarked as Voice Synthesis, Inc. (VSI). Headquarters has nine satellite divisions. On the top of the graphic is the Voice Synthesis Division. This is key because in essence we want to employ AI-based voice synthesis to alert the user to turn off the device and return to the physical world. So, the new corporation has the following integrated divisions:

- Advanced AI in conjunction with IBM's Research Laboratory in Israel
- Advanced Voice Systems with Amazon's research groups

- Delivery Systems & Devices division with multiple vendors for smart eye glasses, lapel pins and other convenient data delivery devices
- Flip Phone Development division centered in Taiwan to assure uninterrupted production
- Network Security division to assure WiFi privacy, use of the cloud and the neutralization of malware
- Quality Assurance division for standards, publications, patent submissions in addition to required inspections
- Education & Therapy division focuses on detox solutions in all venues: home, school and work
- Government Liaison division to enforce the use of Flip Phone devices at all Federal and State entities. This includes taxation, exchange and recycling programs. The division will absorb large numbers of employees who are now "on the street" due to the recent closure projects headed by Cole and Dawn
- Shadow division which is secret and delivers debilitating cyber codes

And I warn you, (as Matt looked directly at Cole and Dawn) I may ask some of you to take an active leadership role in either of these last two divisions."

Cole looked at Dawn, and both nodded their understanding of the potential role(s).

Matt then asked for questions and comments. There were several, primarily focused on the probability of success given the enormity of the national crisis. His response was cryptic and powerful: "If not us, who? Everyone else can't see the problem or has a vested interest in using and/or promoting mobile S-M."

Everyone nodded their heads in agreement.

Matt smiled and said, "That's enough deep thought for a while. Now please join me for a ride in my Wally boat. We'll meet in the foyer in twenty-minutes. And, thanks for your participation in these discussions. Over time, the nation will/must realize that the goal is quite worthy—save the American democracy from Chinese totalitarianism.

The family Maybachs took the visitors to the Key Biscayne Yacht Club. They walked out to the end of the pier and several gasped, Oh, My God!. They saw the 118-foot Wally, a black stealth fighter style jet boat. The guests were then each harnessed in a seat in the raised galley for perfect visibility. The three Rolls Royce turbojet engines soon lifted the bow out of the water to cruise at channel speed to the open ocean. Once there, the 16,000 horsepower roared, and the boat shot forth like a cannon ball and was cruising at 70 MPH in less than two minutes. A couple of the guests compared the feeling to being on the steepest roller coaster when it dropped down. Only twenty minutes later, the craft was floating on the Key Largo Marine Sanctuary. Water periscopes enabled the group to take turns looking at the schools of sting rays and rainbow-colored fish on the reef. Several team members pelted Matt about his $26 million go-fast boat. Murray, who was an accomplished cigarette boat racer, added details about how to handle the craft at high speed. Much to their delight, four members of the team took turns at the helm on the return trip to the yacht club. As the new powerboat enthusiasts were driven back to the Flynn compound, one member remarked just how enjoyable the water thrill ride was, complete without a single "ping" on his cell phone.

Dinner was never to be forgotten by the guests. The 6-courses, complete with fine wines, were enjoyed while the table was abuzz with conversation about the topic which brought them to Key Biscayne. Toward the end of the meal, Matt stood up, made a short toast to America (the land that made the venue possible) and asked the guests to follow him into the library.

With cordials in hand, the friends were eager to hear what the host had to say.

"By now, it should be a surprise to no one that I'm serious about the need and roles of VSI. I'm equally sure that such an endeavor requires smart, focused and honest leaders. That, my friends, is why you're here. Here's what Heather and I have in mind:

- Corporate headquarters will be in Alexandria, Virginia. The location allows the company to apply the talent base of cyber, network security and voice synthesis systems at Amazon only 2 miles away. We will be able to take our pick of the fired Govies. And,

I'm happy to report that Alex has agreed to serve as President and CEO."

(Hear, Hear and Bravo went around the room.)

- A satellite office will be in Bethesda at IBM to coordinate our efforts with the gurus in Israel who apply Artificial Intelligence to voice systems. This work will also entail a new VSI laboratory at the Amazon Headquarters in Arlington. Our new proprietary version of Alexa, trade named "FLIPA" will become smart, very smart, in order to properly interact with the user and warn about the use of a mobile smart device.
- The Delivery Systems division will be based in Memphis across from the Fedex Headquarters. Because they are both pilots, I've asked Murray and Maggie to relocate there for a three-year commitment.

(Glasses were raised to salute the couple.)

- The Education and Therapy division will be headquartered in Boston, conveniently located at MIT where Tinshin will play a lead role in formulating digital detox regimes.

(Approval smiles went around the library.)

- Key chip and device production facilities will be in South Korea, Singapore and Manila. Production will be voice-based alternatives such as Flip Phones, multifunction eye glasses and smart lapel buttons.

- Based in Hong Kong, Chin Chin and Andrew will use their in-country networks to surveil the Communist activities. They have proven to be the most accurate source of intelligence for the U.S. Government over the last two decades.

(Another round of appreciative toasts took place.)

- The cyber guru couple of Cole and Dawn will be valued leaders of our security and SHADOW divisions with real-time visibility into our Amazon, IBM and Government activities. (Hear,Hear!)

Here in Florida, Heather and I will focus on researching and developing incentive programs to reduce the number of smart phones in circulation. And from our home in Hawaii, we will serve as liaisons with governments in Asia. We'll coordinate funding requirements as well as the composition of the Board of Directors. Any questions my friends and new business partners?

There were several questions which all boiled down to: "When do we start?"

"The corporate office will be ready at the end of the month. Heather will be there to help with the furnishing and hiring. Plan on repeating this meeting there in five-weeks. We will detail our three-year growth plan, complete with goals and incentives.

So, with that, Heather and I express our heartfelt gratitude for your contributions to VSI. Please continue to enjoy the beverages and Cuban cigars. All of your transportation arrangements for tomorrow morning have been completed. Thank you again, Gob Bless and good night."

At 11:00 PM, three couples had a midnight swim au natural. The six nude bodies formed a circle in the middle of the pool and raised their joined arms and circled as a waterborne carousel in honor of their esteemed host who was able to "circle the wagons" to create VSI, hopefully not too late to recover enough youth to lead America in the future.

3

ALEXANDRIA

Matt and Heather deplaned their Gulfstream 700ER personal jet at the Reagan Washington National Airport halfway between the District of Columbia and Alexandria, Virginia. The 5-minute limousine ride took them to the Braddock Metro Center Office Building. The elevator took them to the 6th floor where the doors opened into the new lobby of VSI.

"Good morning, bosses," was President Murray's cheerful greeting as the pair entered the lobby.

"Good morning, Alex," said Matt as he looked around the room. "Classy but not overdone. Good job, my friend."

"Thanks, Matt. And thanks to Heather who helped with the furnishings." Heather smiled.

"This way to the conference room. I'm eager to introduce some of our team members, and, of course, they're eager to meet you two!"

The couple followed the president down the hall to the conference room. As they entered the room, eight people stood up waiting to be introduced and greet the couple. President Murray introduced each person and gave a little bio about them and their role in the new company. Soon the party was reseated as Matt and Heather sat opposite Alex at the end of a long conference table.

"Welcome to VSI, my new friends. You have joined a unique company in America. It never had to exist before. Now, God-willing, it will make a positive contribution toward saving our citizens from themselves. Heather and I consider this effort to be primary during our sunset years. America is on the tipping point toward a Socialist government, which you will learn in your new roles, is part of China's goal of global dominance. It's a plan which is being implemented slowly but surely. Few Americans, even most in the current Administration, don't realize it's happening. Over the next hour, I'm going to discuss:

- Poison Media
- Election Process
- Work Ethic

It's because of these subjects and more that VSI exists. I congratulate each of you for your part in our efforts to spotlight, and in some measure solve, the problems. How does it feel to be on the frontier where people will shoot arrows at you?

(A rolling chuckle went around the room.)

Today in America every major newspaper is a bullhorn for the Democratic Party. There is no self-censorship. The editors spit on conservatives, a talent they perfected in college. We'll get to that subject later. The bottom line is that capitalism stands in the way of a Socialist economy, and anyone who says so is branded as a "racist." It uses fake news to destroy the competition. And, as perhaps you've noticed, it supports programs and wide-variety of efforts to dumb down our children in order to create compliant future generations which welcome the central government as the Savior.

So, you ask yourself: "How did this happen? Well, there are several factors. Probably the leading one is that the newspaper's staff is a constant flow of ultra-liberal college graduates. They come from colleges and universities where only liberal ideas are tolerated. The curriculum is full of "social equity" and not "how to earn a living." Journalism schools screen conservative ideas. These liberal reporters do their slanted stories, and the editors choose not to challenge the assumptions and/or facts. And the journalists become expert in what not to report, such as a successful conservative program.

So, why do I mention this? The newspaper employees are S-M junkies always on their smart phones on Apps where Big Tech tells them what to report. The result is that all of the media—digital and printed— is 88% liberal. Many of these journalists seek the security of government employment in D.C. The city is a swinging door promoting these "spin masters" to justify new programs for the "underserved" in order to spend hundreds of billions of other people's money, often in direct violation of the Constitution.

Any questions?

There was only mum silence as the logic set in.

So, how does this media influence our democracy? The simple answer is in many ways at high speed with immediate results. Here are some examples: (Matt counted on his fingers.)

- One falsehood sinks a candidate; it diverts the person from the campaign trail, and there isn't enough time for a comeback
- One ugly picture can infect millions of minds (Hillary Clinton!)
- Unverified quotes and charges of racism
- Negative family and/or friend's interviews
- Financial position (Too much money; too little money)
- Voting Record
- Size of opponent's rallies
- Bullhorns to rally assistance by the unions
- Controversial advertisements and/or celebrity endorsements
- Election Day coverage where unverified vote counts sway voters in Western states

The Bottom Line: Conservative candidates lose, and the swamp and the deep state grows.

So, my friends, you see the essence of democracy, the election process, is corrupted for political and personal gain. The seeds of a totalitarian regime grow into trees creating a forest of deceit. So, how do you combat this? Well, part of the answer is with 10 year-old students on a school bus. It will become much clearer after tomorrow. Time for a coffee break!"

"So, a corrupt media can dictate the results of an election. And just as bad, perhaps even worse, it can foster a lethal weapon against our republic: deteriorated work ethic. Don't believe me? Here are some revealing statistics:

- 50% of the population hasn't had a job before age 25

(There are two jobs for every unemployed person.)

- Only 30% of the unemployed are actually seeking work
- Over 67% of high school and college graduates have no useful skills in the economy
- There's little or no inspiration from family and friends to find work
- There aren't even recruits for the military

But it gets worse! Why?

Many of the youths don't need a job. Did you know that a family of four (2 children) gets $72,000 annually in unemployment benefits? PLUS:

- Energy subsidies
- School breakfasts and lunches
- Food Stamps
- Obamacare
- Housing Vouchers

And there are BOTH federal and state programs for the above. Therefore, the need to seek work disappears. And, you guessed it. S-M reinforces this "quiet quitting." "The government will take care of us until we get our inheritances."

So, what's the result? 70% of the items at Walmart are made in China. 96% of all S-M devices are made there. And, as you know, these devices have thousands of time-consuming applications which are more "valuable" than going to a job. And what really makes the situation so tragic is that the 10 year-old on the bus is told that China is our friend by:

- Parents
- Teachers
- Newspapers
- Television
- Posters
- Commercials
- Hollywood
- Sports figures

So, a democratic republic dies one child at a time. Let me repeat: VSI's purpose in life is to be a lighthouse that shines on the problem and offers effective, lasting solutions. Thank you for your time and welcome aboard."

Everyone clapped and shook Matt's and Heater's hands as they "went back to work!"

At 2:30. The Padlockers, Cole and Dawn Langford, arrived at the offices. The infamous couple in Government circles was hired to lead the security and cyber weapon divisions of the company. The couple spend an hour briefing the staff about network security. It was a give-and-take session with lessons learned which would serve VSI well in the future as the company grew and became a target of cybercriminals.

Afterwards, the couple joined Matt on a park bench just outside the building. Dawn was the first to speak. "Matt, Murray and Maggie will face an extremely hostile audience tomorrow based upon Facebook chatter."

"I'm sure of that, Dawn. We had to send a synopsis of the presentation in advance. I didn't think it would be approved. But it's a phony attempt to appear non-biased. We will use the video for marketing. So, what's happening in your SHADOW world?

"Well, we have some real progress to report. Cole, please tell him,"

"Matt, we've made progress thanks to the great employees at our Padlockers headquarters in Bermuda. The staff does a frequent mind-meld, and the results are incredible. Within a month, the testing of two new products will be completed. "DigiBomb" renders a device useless forever. And "NetBomb" ends digital and wireless transactions. Period. We are able to create rolling black outs via code replication which is invisible and can not be detected my modern means. The target, say a power station control console, never knows the source. It becomes forever useless until we re-activate it after achieving concessions and/or cash from the owner. IBM's AI Laboratory has really helped us develop and test these products. Oh, and one thing. We are not going to submit patent applications so the Chinese can't reverse-engineer our designs."

"Guys, that's terrific. Thank you. By the way, we'll have dinner at the Morrison House tonight. Heather and I would like to show our appreciation for all that you've done for the country and now VSI.:

"Boss, love to. We'll be there."

The management team of Matt and Heather, Murray and Maggie, Alex and his wife and the Padlockers enjoyed Happy Hour in the petite lounge at the Morrison House boutique hotel in the city center. Dinner was at one end of the small dining room. It was calm and classy; so appropriate for the occasion. As the afterdinner drinks arrived back in the lounge, Matt asked Heather to pull an envelope out of her purse.

"Cole and Dawn, we've always enjoyed visiting your office and condo in Bermuda. You were right. It has attracted the key cyber talent on the planet necessary to create your products. Your initial applications have disrupted supply lines and military preparedness of America's adversaries—all without the Administration's knowledge.

You know that large condominium above yours at Savoir?"

The couple answered "Yes" in unison.

"Here's the deed. You own it now, complete with its 180- degree view of the Bermuda paradise."

The Murray's clapped. Everyone hugged each other. A bottle of champagne topped the evening. Matt couldn't resist the opportunity to offer a toast. "What's going to happen in Asia?

Only the SHADOW knows!"

4

GEORGETOWN

"Good Morning ladies and gentlemen and welcome to Georgetown University's Malloy Lecture Series. I'm your hostess, Cheri Valentine. Today's topic is both revealing and controversial. The Flynns, Murray and Maggie, have been on the forefront of the anti-mobile social media movement. You may have read one or more of their recent articles. If so, you are armed and dangerous with questions for our guests. Please hold your applause, gasps and/or comments until the end. So, please welcome to the stage Murray and Maggie Flynn with their topic: "Social Media creates Social Mummies."

(Impolite small applause and snickers filled the auditorium.)

"Thank you, Miss Valentine. We're excited to share our research with such a knowledgeable audience. Over the next 50-minutes, we will describe and show videos of one-half-dozen common situations in our society. We expect our observations to make you uncomfortable. In fact, if you don't squirm in your seat, we've failed in our mission. Why? Because you represent the very pinnacle of social media use in America. On average, you've used social media for twelve years already. And the odds are excellent that you each have smart phones, smart watches, tablets and laptops to ensure total connectivity with the world.

So, as a courtesy, and so you'll get the most out of our remarks, we're asking you to take out your smart phone, turn it off and pass it to the person on your left. Don't Worry! You'll get it back after our presentation. But to fully appreciate the impact of these devices on our society, you need to live without a "ping" or vibration while we show you what's happening in our country and WHY it's happening."

As Maggie stepped forward, adjusted the microphone and pointed her laser pointer at the large screen behind her, a woman in the audience screamed, "I can't do it. I can't do it!"

"Ladies and gentlemen, I direct your attention to the first of six very short videos."

Maggie narrated a video recently taken at Brown University. In it, Silvie and Krystal sat next to each other in a Chemistry 201 lecture. Both women were considering medicine as a career. They were both fluent in a wide variety of digital devices. During the lecture, Silvie took notes on her laptop while listening to music in earbuds and texting her boyfriend at crew practice.

At the same time, Krystal spend most of the lecture programming her new smart watch while recording the presentation. She used her smart phone to take pictures of several of the key slides presented by the professor. She thought to herself: "Why waste time here; I'll just set up a tape recorder and pick it up later."

After class, the friends walked to the student union and smiled to friends while being on-line in their heads. Krystal congratulated herself on being able to get her email on her new watch while simultaneously texting videos of the goal she scored in a recent field hockey game. As they reached the union, they were greeted by two sorority sisters who were leaving for a political demonstration which had been broadcast on the campus network. It was a protest against a conservative giving a speech about runaway inflation. Several of the sisters showed up with protest signs made just minutes ago via a text alert. On the other side of the campus Green was a conservative speaker questioning the value of a liberal arts college education. An hour later, they got a text about a "Post Rally Brew" at a local pub. The sun was setting behind a deserted library, as the two friends walked past on their way to Mulligan's. They were too busy responding to text messages to talk with each other.

At the pub, they met other sorority sisters who joined them in reliving the recent "bashing" of a guest speaker on campus. They each looked at their smart phones to view videos of the event. It was clear that this group, one of hundreds, was used to using social media to voice their opinion and ultimately dominate the political discourse on campus and in the state of Rhode Island.

At the end of the video, Maggie turned to the audience and asked, "How many of you see yourself in this video? Please raise your hand."

Almost everyone in the audience raised their hand.

"I thank you for your attention. Now Murray will show you where the young women in the video probably met the next morning....STARBUCKS...where the environment effectively mummifies the patrons to silence."

"Thanks, Maggie. Yes, from this video graphic, you can visualize the silence. Indeed, look carefully. Do you see yourself in one of the following positions? (Murray points to the coffee bistro layout.)

- Barista working two machines while looking at her smart watch and listening to music via earbuds
- Server #1 working the cash register while serving pastries and texting her girlfriend sitting ten feet away
- Server #2 serving pastries while changing the channel on the videowall
- Patron, in line readies an electronic credit card
- Patron, in line talks to girlfriend at another STARBUCKS
- Patron, in line takes an order from her office
- Patron, in line turns up her heated seats in her automobile
- Patron at a table plays fantasy football on a laptop
- Patron at a table works on a homework assignment
- Patron at a table watches a soap opera while consuming 4 cups of coffee
- Patron on smart phone learning English as a second language
- Patron ordering a poster: "Free Meth for Mankind"

- Manager uses tablet to coordinate the next day's deliveries and change the outdoor digital marque sign with new specials
- Nobody looks up from their devices. Person-to-person eye-contact is to be avoided at all times because it might lead to a diversionary conversation. There is no sound. It would distract from the mobile S-M devices.

Murray then brought up the next video which showed the inside of a #7 subway car enroute from Queens to Grand Central Station in New York City.

"As you can see, the scene is almost identical to STARBUCKS. Seating and standing, people are sipping coffee while looking down at their smart devices. A quick scan showed:

(A smart phone rings in the audience. Murray pauses and smiles.)

- A young couple holding hands, each in their own earbud world
- Mother between two youths on GameBoys
- Junior stockbroker looking at his smart watch and shaking his head
- Junior college student typing on her laptop while listening to acid rap music
- Young man with hair braids down to his buttocks watching an NBA video
- Old woman using the smart phone to translate Hungarian

- Two teenage girls texting each other
- Young priest with earbuds and his lips moving to organ music
- Father and son in baseball uniforms looking at a video

At the same time, on the advertising marque above the windows were streaming ads for cleaning services, home safety systems and forthcoming electronics trade shows

As the train stops, riders still looking at their smart phones exit the train while others enter, also looking at their devices

A low rumbling sound was heard like a wave in the audience. It was clear to Murray that this presentation was working. The college students were very uncomfortable away from and were, indeed, addicted to their mobile smart devices.

Murray handed the microphone back to Maggie who introduced the next scenario by saying, "Please raise your hand if you've ever gone camping and slept overnight in a tent." About one-third of the audience raised their hands. Many looked around to see who was raising his or her hand. The next video was started showing a small family around a campfire next to a raging river in Yellowstone National Park.

As the father was starting a fire, his wife said, "Honey, I'm so glad we did this. I needed to get away from all those devices in my world!" As he readied marshmallows on long sticks, he quickly looked down at this smart watch to check his stock portfolio. A moment later he checked his video doorbell and lit lights at the front door. The mother sent a text to her parents: "We're in Mother Nature; pictures in the morning."

The hamburger dinner tasted better in the clear air of the Sierras. Smores and toasted marshmallows reminded the couple of their childhood. At 8:30 PM the family members checked their lanterns and went to their respective tents.

- TENT 1. The parents watched a streaming video on a device propped up on a folding camp stool at the end of their sleeping bag. After the movie, they watched a short video of their dog at the kennels. They kissed and survived the cold night in each other's grasp.
- TENT 2. Two young teen boys watched a video of an ice hockey game. One sent a text to his girlfriend in El Segundo. She answered with a text displayed on a smart watch. One charged his laptop for homework in the car on the way back to Los Angeles.
- TENT 3. Two young girls sent videos of them wading in the cold mountain streams to their boyfriends. They watched an x-rated video before turning off their Coleman lantern.

The audience buzzed about the video as Maggie turned to it and asked: "Do you see yourself in this video? Could you go camping without a smart phone?"

The audience's response was immediate: "HELL NO!!"

As Maggie handed the microphone back to Murray, she thought to herself: "This smart phone addiction is much more rampant than the polls indicate, and, therefore, VSI's work is even more urgent."

"Now let's go from a national park to a boardroom in Manhattan. You're looking at a conference table ringed by a dozen executives and a corporate secretary. A rotating 3-D hologram spreadsheet showing the impact of the company's latest acquisition is projected above the middle of the table. "Let's see how these people are paying attention."

- VP, Finance. Confirmed budget meetings via a smart watch
- VP, Operations. Accessed E-bay for special equipment sales
- VP, New Business. Listened to Beethoven in his earbuds
- VP, Legal. Viewed videos of a daughter in daycare
- VP, Plans. Used a laptop to research Wall Street events which could impact the acquisition
- VP, HR. Used smart phone at access her new vehicle's maintenance agreement to determine the meaning of an idiot light on the dashboard. In the background, she scheduled new hire interviews.
- VP, Logistics. Used smart phone to determine if a railroad strike was eminent.

The secretary accessed a clandestine WiFi monitoring application to determine that during the meeting, occupants:

- Made 44 Smart phone calls
- Accessed Google 31 times
- Watched 6 videos
- Purchased 4 items on E-Bay
- Wrote 33 texts via smart phone and smart watch

As the video ended, Maggie made a couple comments. "So, you don't have to learn math equations and cursive writing. The devices do it all. And you should understand that the data from all of the scenarios is able to end up in Shanghai."

As she started to hand the microphone back to Murray, several audience members shouted obscenities and walked out of the auditorium.

Murray asked the audience to be patient. "Ladies and gentlemen, our goal is point out the impact of these devices on our country. You are all involved. And you are free to retrieve your device from your seat neighbor and leave. However, we only have one more video to show. It relates to all of you."

Murray started the video showing elementary students on a school bus.

- Amy. Studies her notes with music earbuds
- Leroy. Plays basketball on GameBoy
- Lexi. Uses the bus WiFi to access the Internet for homework answers

- Tanner. Uses the smart phone to solve Rubic's Cube
- Calder. Watches last night's World Series game
- Sophie. Uses her S-M to bully a female classmate
- Anthony. Dozes off to new country rap music
- Ryan. On the smart phone to Tanner about an upcoming field trip
- Brandon. Rehearses band music
- Jason. Reviews football formations for the "Sandlot" league
- Emilie. Watches a video about her dog at a dog park
- Tiffany. Talks to her mother in a trailing SUV who is bringing a forgotten lunch
- Driver. Uses smart device to show the bus location at the school's central command center.

"So, let's sum up the impact of the devices and why you, and all your friends, are turning into socially-dead mummies!

More people shouted and walked out of the auditorium. Murray counted on his fingers:

- Devices are made by Foxcomm in Zhengzhou, China. We lost the innovation and manufacturing
- Devices eliminate your conversations; essential for business success
- Devices train you to automatically look up something rather than think something through
- Devices create Dumb Downed youth ready to follow the dictates of a central government
- Devices make eye-contact almost illegal

Now please retrieve your smart devices. How many of you immediately turned on your device for a contact purpose?"

All arms in the auditorium were proudly raised in protest!

"Guess what? Perhaps you should consider carrying only a Flip Phone as a security device. You'd be surprised just how clearly you can think without omnipresent digital distractions."

The audience erupted in a thunderous chorus of boos and profanity. A few stormed the stage to voice their disapproval of the suggestion. The moderator signaled for security guards to escort Murray and Maggie from the building and to their car for a safe departure. Matt and Heather quietly left though a rear door.

A half-hour later, the foursome met in the lounge at the Morrison House for a "I told you so" session. Murray took a sip of his whiskey and remarked: "I knew it was bad but not that bad, Dad!"

"Son, that's why we're here in Alexandria, and why VSI must be successful."

5

ARLINGTON

Promptly at 8:00 AM, the Flynns (Matt, Heather, Murray and Maggie) signed in at front desk of Amazon's East Coast Headquarters at the Potomac Yards only a 5-minute car ride from VSI.

"Good morning, Flynns. I'll tell Alan you're here."

Moments later the foursome was introduced to the senior staff of the Voice Synthesis division in a sixth-floor conference room. This division is composed of computer gurus from the West Coast who invented, improved and maintain the world-changing voice synthesis application, Alexa. Introductions and fresh cups of coffee helped initiate a candid and productive meeting.

"Alan, thank you for seeing us on such short notice. And thank-you for your cryptic notes on the White Paper I provided last week. I hope you can tell we're VERY serious about working with you to leverage your technologies to solve what we view as a critical national problem."

"Matt, your family's contributions to this country are legendary. When you have an idea, the world listens!"

"Thank you. You are most kind. As you know, we've created Voice Synthesis, Inc. (VSI), a non-profit corporation just three miles from here in Alexandria. We want to create joint project teams to develop our line of products which should result in a very lucrative revenue stream for Amazon. I'm probably saying what you already believe: the world's social media and computer system access will be dominated by voice commands and voice requests, ideally done by Alexa. As our White paper demonstrated, there is a fundamental national movement to abandon the keyboard on social media devices. And the devices preclude eye-contact with the rest of the world. The user withdraws into La-La Land and does not create and/or learn products or services essential for robust economy within a democracy."

"And it doesn't help that all of the devices are made in China."

"Precisely!"

"Yesterday, Murray and Maggie gave a great presentation about the problems inherent in social media at Georgetown University. In the end the audience was urged to give up their smart phone in favor of a Flip Phone. The audience booed, and they had to escorted off the campus. This event demonstrated the depth of the national addiction, and it confirmed for us that a new technology must be focused on the pre-teens in order to be successful."

"Sir, that's one reason why we have an entire Alexa product line for Amazon Kids."

"Yes, I know. In fact, the Alexa applications for children are incredibly useful. Your team is to be congratulated. Heather and I counted up about two-dozen services which .

a youth can access with a verbal command. They include but are not limited to:

- Practice Math
- Learn Spelling
- Learn Reading
- Daily Routines
- Audible books
- Alarms; such as medication time
- Entertainment
- Emergency numbers

Amazon cracked the code and determined that the future of application access was via voice. We're here today to leverage this finding with new products for the youth market. We must break the link between the finger and the keyboard. In effect, we wish to apply your "Sticky Notes" and "Routine Profile" in order to increase a ten-year-old's productivity without being overtly distracted by a mobile S-M device. We desire to do this to prevent our society from being Dumb Downed to a point where there will be no rebellion against the Asian Dragon."

"Mr. Flynn, I've never heard it so aptly described. We sell S-M devices of all kinds. But my management feels the same way you do. It's not enough to provide voice commands such as "Alexa, play Roy Orbison." We must take the commands to the next level: urging and teaching skills so the youth can escape from "La-La Land" as you call it. What do you want us to do in this joint project you mentioned?'

"Over the next six months, I will fund a joint effort to develop voice services using the access command: "FLIPA." This is far enough away from Alexa so as to avoid confusion. FLIPA will be solely used on new access devices which we jointly develop like eyeglasses, lapel buttons and, of course, Flip Phones. No Internet, No keyboards. No movies. Nothing is available to divert the user/wearer from productive tasks. VSI will execute incentive programs for consumers, schools, hotels and hospitals to buy our products made in Texas. Families will pay a very small monthly fee for FLIPA. When a youngster completes a number of student routines, learns a new skill like cursive writing, award points will be redeemed for Amazon services or retail purchases at Barnes & Noble which needs the traffic to stay in business."

"Facinating, Sir. Let me see if I understand you correctly. You plan to create new FLIPA apps which leverage Alexa's capabilities for the purpose of focusing youth on voice-activated systems and away from the in-your-face social media device. This will sell more Amazon products which we jointly develop. And monies we devote to these projects will be tax write-offs for donations to a non-profit corporation."

"Alan, I couldn't have said it better."

The executive saw a lot of nodding heads around the conference table, smiled, reached across the table to shake Matt's hand and said: "You've got a deal."

6

DISTRICT OF COLUMBIA

Murray came to the District for an important meeting affecting the future of VSI. In a short taxi ride from Alexandria to the Pentagon, Murray said to Alex, "We've submitted draft documents and White Papers per request. These people are now aware of our plans and will have a lot of questions. Our logic should prevail, but we're dealing with government agencies with hidden political agendas. We'll see, but one certainty is for sure is the impact the Padlockers will have during the demonstration.

Cole and Dawn flew in from their corporate headquarters in Bermuda and were waiting at the security checkpoint. The foursome greeted each other and waited only 15-minutes before they were introduced to a group of sour-faced men and women in a 5th floor conference room. Alex knew a couple of the men and guessed that the audience was comprised of "spooks" from the CIA, DIA, NSA, DARPA and the Department of State. Dawn connected her laptop to a projector and nodded her readiness. Alex introduced Murray and the Padlockers and said, "Ladies and gentlemen, we have a 30-minute briefing which we have privately developed and hope that versions of the software modified by your organizations will serve national security requirements. Cole, you have the microphone."

PADLOCKERS

**Novel Civilian Closure Corps (CCC)
reduces Federal Bureaucracy to limit
the Slavery of Socialism**

JASON O'NEIL

"Thank you, Alex. It's an honor to be here. You have our bios and know all about Dawn and me from our three-year project to padlock incompetent and/or unnecessary government departments. We were directly responsible for reducing the NSA and the DoD by 50% with NO reduction in mission effectiveness. We are here in person because the briefing is too sensitive for a video conference, no matter how secure you believe it to be here. Today we will show you the preliminary results of tests we're conducting aimed at shutting down an adversary's computer systems...all of them for good until we release the cyber strangle hold on the operating systems. Please ask questions as they occur to you. Cole nodded to Dawn to start the presentation entitled: "DigiBomb and NetBomb." In a series of five short scenarios, Cole explained why the control consoles went dark. The video showed operators frantically attempting to restart their machines to no avail. One video discussed how the cyber-killer software was virtually undetectable through the use of Artificial Intelligence provided by the team's partner, IBM.

"It would take multiple supercomputers operating simultaneously to figure out how the software works, much less stop it. There's no known way to recover unless WE do it. This is war in the 22nd Century here today. All of the battleships and rockets are useless once our DigiBomb is activated.

Incredulous glances went around the room.

At this point Cole nodded to his wife to remove a thumb drive from her laptop. He walked over to her, put out his hand and accepted it and said, "I hold a force many-times more powerful than a hydrogen bomb. He handed it to the host, the Assistant Secretary of Defense, General Taylor, who then passed it around the conference table. The Secretary of State asked, "Why did you develop it? Bermuda is not target."

Cole was quick to answer, "Mr. Coleman, our company, VSI, exists to reduce the deleterious impact of social media devices on our country. The devices come from China. They infect our youth with a negative work ethic in order to make our citizens dependent upon a central government. VSI will initiate a campaign to block the importation of these devices into America. We will petition the Administration to levy extremely high tariffs on Chinese mobile digital devices in order to discourage export to this country. At the same time, we are building a factory in Texas to provide a replacement product, which when properly used, should wean the youth off of the debilitating S-M devices."

"Oh, I see. And what if the Chinese persist in filling the demand here?"

"Sir, you just witnessed the ultimate weapon. From an unknown, untraceable source, a plant(s) which makes the S-M devices will go dark. And it will remain dark until WE tell them what to produce—-access devices which do not dumb-down our youth. Do you understand what I'm saying? was the question Cole asked as the thumb drive was handed back to him.

"Mr. Langford, your technology is rather Earth-shaking. Am I correct that you come here with a request?"

"Yes, Sir. We do. You are the Deep State and rule Washington. Our plan is to have dumb devices, like Flip

Phones, replace smart devices. Millions of people will be incentivized to have their smart devices recycled in mega-landfills. We ask your support to: (Cole counted on his fingers.)

- Provide grants to develop and maintain these landfills
- Provide grants to partially fund two mega-factories to produce the Flip Phones here (We'll use alien laborers who have applied for citizenship and therefore will reduce welfare payments.)
- Provide grants to states to embed the requirement for the use of these devices in schools over union objections

(Alex and Dawn's eyes met with a glint of success in them.)

- Allow us to improve our DigiBomb software as a classified program within the DoD's Cybercommand. Our company shall remain a well-guarded secret from everyone: the public, the Chinese and the President to preclude attribution of any kind."

"Do we have your approval for the grants and the secrecy? We are a commercial, non-profit company which operates on pre-approved schedules. In a perfect world, this august body would inform us of your decision within a fortnight so we can establish joint project teams on a timely basis."

"My new friends from VSI, you will have a decision within that period. As you know, we have special accounts just for such contingency projects. Thank you for your presentation. It's one everyone in this room will never forget."

The VSI employees were given a ride back to the Morrison House in a black Suburban with dark tinted windows. The foursome had some pre-lunch Cava in the lounge as Murray called his parents to report on the meeting.

"Dad, we did exactly as you instructed. Cole was awesome. Based upon the logic of our request, the power of the bomb and the chemistry in the room, this foursome expects a call to commence the project teams within a week. Thanks so much. See you tomorrow. Love ya!"

7

LEESBURG

One week later, Matt flew to Washington to join Murray and Alex in a presentation to a county school board in Leesburg, Virginia 26-miles west of the district. Over the previous month VSI submitted several White Papers about "Teaching to be Smart about Social Media." Two hour-long Zoom sessions were held with the members, several of whom were motivated based upon their experience with their own children. The goal was to get a pilot program for VSI to formulate Congressional legislation as part of a national campaign. The trio took a company Turbopod for the 12-minute flight to Leesburg.

TURBOPOD

NOVEL PERSONAL ELECTRIC AIR VEHICLE
REVOLUTIONIZES GLOBAL TRANSPORTATION

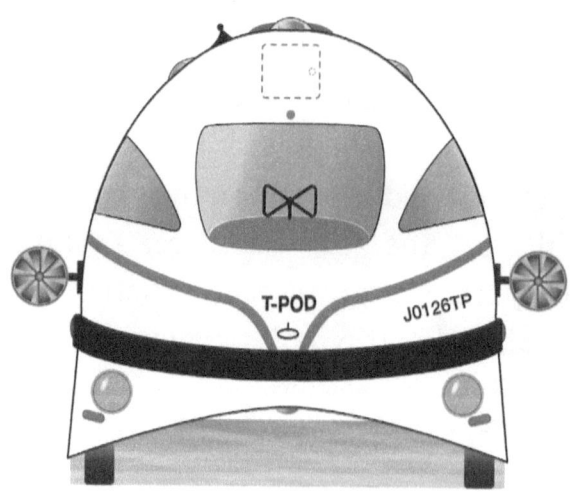

JASON O'NEIL

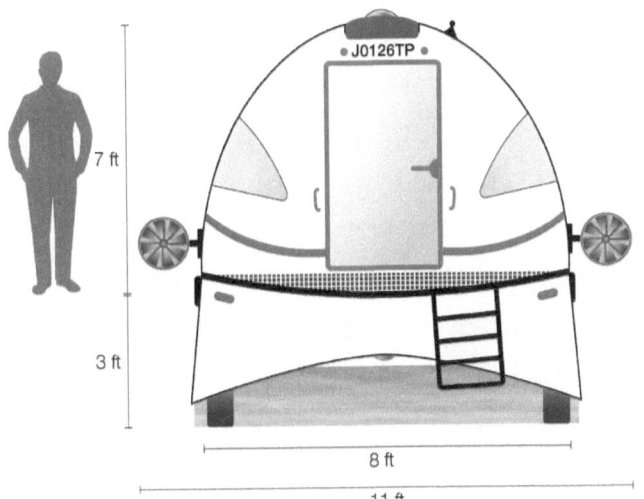

7 ft

3 ft

8 ft

11 ft

J0126TP

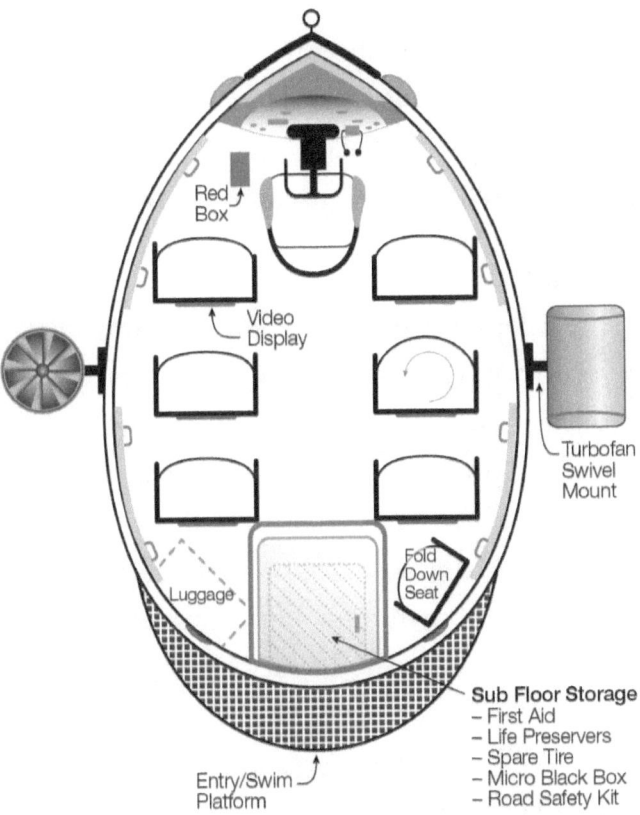

Red Box

Video Display

Turbofan Swivel Mount

Fold Down Seat

Luggage

Entry/Swim Platform

Sub Floor Storage
– First Aid
– Life Preservers
– Spare Tire
– Micro Black Box
– Road Safety Kit

In the Lee County office building, the trio was greeted as sympathetic souls helping to confront national elementary and high school initiatives which embed racism and lies by far-left politicians to further their national control agenda. Matt stepped up to the microphone:

"Thank you for your warm welcome. As you know, we come with a simple message to a complex problem: Save our Children from excessive and corrupt social media which dumbs our youth."

"Members of the Board let me start by saying that there are some good things from S-M. It can help a child:

- Learn English, Math, Civics and Cursive Writing
- Provide a safety whistle
- Access health information
- View Bulletin Boards
- Stay connected to family and friends

But, it's not all good!

Our President, Alexander Murray, will present our proposal for a prototype trial to begin the process of weaning our youth, your sons and daughters, off the addictive devices. Good friends, I present Alexander Murray."

"Thank you, Matt, and thank you for this opportunity. As you know, during our Zoom sessions, we highlighted the damage the continuous, omnipresent mobile social media does to our society. In large measure, it is directly responsible for the lack of credible leaders in America. How can that be? Well, beyond the good things, the case against prolonged S-M is pretty clear:

- Promotes superficial thinking
- Eliminates eye contact
- Demands expensive technology updates
- Foments homelife disobedience
- Promotes anti-democratic philosophies
- Promotes explicit sex at an inappropriate age
- Promotes streaming video, a distraction from homework
- Leads to anxiety, fear and depression
- That's a partial list. It could also include real-world hurtful bullying.

All of the above impacts deteriorate the national work ethic which must be maintained in order to preserve our democratic republic.

We're all parents responsible for the appropriate use of S-M during the day. It can't be a digital babysitter. Parents must educate their children to think twice before hitting the "enter" button. And whatever is typed must be able to be seen by parents, teachers, clergy, grandma and future bosses.

Several board members nodded their heads in agreement.

Our proposal is as basic as the Flip Phone itself. We want to make it an important tool in the battle against unchecked S-M. As an example of the ugliness, I will turn the microphone over to my colleague, Murray Flynn, who will briefly describe the national problem of bullying by telephone."

"Thank you, Alex. Fortunately, my wife, Maggie, and I were able to rescue our teenage daughter from cyberbullying. It took six months, but we did it. Our research has uncovered that 34% of teens between the ages of 11-15 have experienced cyberbullying. And it's growing at 12% per year. In many cases it leads to depression and absence from school. On a national basis, this borders on an epidemic tragic toll on humans who otherwise would be able to enjoy the American experience. Please refer to Page 16 of our most recent White Paper for a list of symptoms:

- Suddenly stops using a mobile S-M device
- Uneasy about going to school
- Jumpy, nervous when on S-M
- Loses interest in favorite things
- Doesn't want to talk about it
- Avoids friends
- Drastic changes in eating and sleeping
- Passing statements about suicide

The answer is simple: Teach Youth to Never Respond to Cyberbullying. An answer VSI suggests is baseline the Flip Phone. It can't access the Internet, run videos and perform functions essential for cyberbullying. This is controversial, but our research proves it works. That's why we propose a trial with a 7th grade in one of your schools. The program involves:

- Flip phone for each student with our proprietary FLIPA voice synthesis to remind the student about performing actions required by the school and the student's personal profile
- Rack of cubbies at the entrance: secure and powers the devices during the school day
- Instruction about its rationale and use

- Summary Report after a two-month trial. And we can provide a transaction history per phone per day per student, if requested. Indeed, you can independently decide if the program works. We think you'll be pleased with the results as the youngsters actually:
- Speak in public
- Make eye contact
- Use the library for research
- Are more appreciative of others
- Take part in classroom discussions
- Have more time for homework
- Have better grades

And that, my friends, is why we at VSI are so eager to help your students. The Flip Phone actually forces the individual to focus on life, not a La-La Land which distorts opinions and diverts attention for gaining the talents essential to participate in a democracy. I welcome questions."

Murray adroitly answered questions about the phones, cubbies and data reports available to the Board.

Promptly at the hour mark, the Chairperson stood up and stated: "Gentlemen, thank you for your most informative presentation. You will have answer about the trial in ten business days."

As the trio was walking across the parking lot to their Turbopod, high-five salutes celebrated the occasion. As Matt got into the craft, he said: "Guys, the truth is on our side. We will prevail. These people know what's happening to their children, and our presentation will spur them into action."

8

NEW BRAUNFELS

One week after visiting Virginia, Matt's Gulfstream landed at the small airport serving New Braunfels, Texas. Located sixteen miles northwest of San Antonio, this river-rich area was settled by the Germans in the mid-1880's. As a county seat, the city has attracted both small and large industry. And it's a year-around tourist destination with its shallow flowing rivers ideal for innertube regattas and float parties.

Matt met Alex in the corporate terminal about two-miles east of the city square. They rented a car and drove to a white wooden two-story farmhouse of the 300-acre Reneu Farm. At the dining room table the Chairman signed the purchase agreement under the watchful eyes of two attorneys from San Antonio. Smiles went around the table as the previous farmer-owner took Alex to his antique barn just outside the kitchen door. Meanwhile, Matt called the President and CEO of Alcatel in Paris, France, the world's leading Flip Phone manufacturer.

"Maurice, it's a go. Please have your site survey team here as soon as possible to commence the factory design project. Our lawyers here will obtain all the appropriate permits. And please plan to be my guest in Key Biscayne next month."

Matt smiled as he closed his own Flip Phone with a shamrock image on the outside.

After lunch at the Bierstube on the square in town, the duo motored south eight-miles to one of Amazon's largest Fulfillment Centers in Schertz, Texas. The center manager was proud to tour the VSI executives through "his" one-million square-foot baby. The endless conveyors and moving packages really impressed the visitors. They congratulated themselves on their decision to produce the phone in Texas, so close to a major distribution point. It was also important that Amazon had another similar center just north of New Braunfels on a different power grid.

At 4:00 the executives drove to San Antonio for a dinner meeting with executives from the Blue Signal Staffing Agency. This company is the largest recruiter in the region. It specializes in vetting newly-arrived aliens who have applied for citizenship. Most have some English language skills, and all have a stellar work ethic to earn money to support a family. Perry's Steakhouse on Cantera Parkway proved to be the ideal place to initiate a long-term recruitment plan. "Our people will get with your people to execute the paperwork," was Matt's parting comment.

At 10:00 PM the Gulfstream took off for the long flight to Singapore via Hawaii. In flight, Alex received a text: "Amazon-Alexa is a go!" Only twenty minutes later over the Pacific Ocean, Alex received a second text: "Leesburg is a go!" The two smiled and clinked their glasses of Blanton bourbon as Alex was quick to say, "Boss, our Irish luck prevails. Erin go bragh!"

9

SINGPORE

The Gulfstream landed just before noon, and Matt and Alex were driven to the famous Raffles Hotel for lunch. During the meal in the ornate 130-year-old colonial restaurant, Alex confirmed the afternoon meeting with the Flynn family friends from Hong Kong. He then excused himself to confirm that the discussion area in the courtyard was ready.

Chin Chin Po and Andrew Ho, both American citizens, were parttime, unofficial covert officers of the CIA. Over two decades, they had developed an extensive network of informants throughout Mainland China and Taiwan. Because the couple could be observed by Communist Party operatives, the proper setting was imperative for discussions about China.

Alex reviewed the setting with a 20X20 foot square area with shoji screens for walls. To thwart eavesdropping, white noise generators were suspended from a tree to drive the sound down into the square. Music speakers ringed the square. Alex sent a text to Matt: "Ready Boss."

Right on schedule, the couple entered the hotel and were escorted to Matt's table. Warm greetings and hugs were both sincere and long overdue. They joked about the Wally boat ride, "never to be forgotten."

The trio then walked out to the courtyard and entered the "Quiet Cube" as Alex called it. Cell phones were turned off and put in a metal box. A Tea Ceremony was on the circular table in the middle of the Cube. Matt initiated the conversation.

"My good friends, we've just come from Texas where we purchased a site for the new Flip Phone factory. It should be commissioned in 20 months. It can't come soon enough because it's clear to us that every day thousands of preteens get hooked on the smart phones made in China. So, what is happening there that could impact our planning so we can take prudent steps."

Matt served his guests tea and pastries and motioned to Chin Chin to start the discussion.

"Matt, as you know, the Communists have developed a national on-line Social Credit System or SCS."

'I know only what was briefed by Tinshin in Key Biscayne."

"Elements of the system are already in America, ready to be implemented and expanded when the Chinese assume control of our Republic."

"Wait a minute! Are you saying this social media system could be instrumental in our downfall?"

"Yes. We're convinced of it. And the final nails in the coffin with be nationwide Digital ID's and cryptocurrency. Citizens won't be able to move or spend without government visibility and, ultimately, control. But, first, let me describe the current situation."

"Ok. Please do. I'm all ears!"

"Today in China there are tens of millions of cameras watching hundreds of millions of people, typically in cities. They are everywhere-traffic intersections, terminals, universities and public buildings of all kinds. Behind the lens is a lot of Artificial Intelligence that can identify an individual and know precisely where that person resides, works, plays and just lives!"

"Ok. But we have a lot of cameras in America."

"True. But what makes the Chinese systems so sinister is that it is the front end of a Social Credit System, a system used by the CCP to control the behavior of every person in the country. Indeed, their own propaganda about the impact of this social media is to allow trustworthy to roam everywhere, but make it hard for the discredited to take a single step!"

"Jesus, that's horrible. It's total control of a person's movements, indeed, whole life!"

"Yes, and it's even worse than that. Some of the actions the central government deems as "bad behavior" which warrants punishment include:

- Traffic offenses
- Smoking in public places
- Deadbeats (who don't pay their taxes or bills)
- Making frivolous purchases
- Playing too many videogames
- Watching pornography
- Consuming too much alcohol or junk food
- Criticizing anything about the Government, particularly the social credit system

And here's the real kicker—being friends with or messaging others with low social media scores by committing one or more of the above offenses and many more."

"So, how does one get around this system?"

"There are essentially two ways" 1) Pay bribes and 2) only use a Virtual Private Network (VPN) independent of the spying government....at least for now."

"Oh? Please explain."

"First, bribes are common place used by penalized citizens to gain services. Often these bribes are with hongbao (envelopes filled with cash). Even to get packages out of customs requires hongbao, particularly if you have a low social credit score. And it only takes one offence, and your score is downgraded FOREVER. As of now, there's no way to raise your score, except, perhaps, by bribing a government official in a watchdog agency."

"At this rate, you'll need a bribe to do anything in China, It's like everyone is a spy with their hand out!"

"You're right, Matt. Many use a VPN which allows visits to restricted websites, use of banned apps and bypass apps required by the government. Today, in Beijing you can purchase a modem with a VPN pre-installed, and, of course, it requires a cash bribe paid in a location safe from a prying camera."

"We've talked about how China is infiltrating America. It makes all of the smart devices, owns Hollywood and the national media, cable networks and software companies throughout our country. Tech Giants in Silicon Valley spy on everyone. "Wrong Think" is banned by Twitter, Facebook, YouTube and TikTok which is owned by China.

Criticism of Washington's policies is banned. And America buys hundreds of millions of cameras from China. They are in every portable phone and even doorbells. So, the social credit spy network infrastructure could be harnessed with a few directives from Washington or Beijing."

"You're right. I have more than a dozen cameras around my home, and Heather has a new smart phone with three cameras."

"Matt, and, as I said before, it gets worse!"

"What could possibly be added which would foster this total subjugation of the individual?"

"As I said: Digital ID and Cryptocurrency!"

One's daily actions will be caught on camera and time-stamped. One's digital signature, including genetic material, will be coupled with personal history for algorithms to predict future behavior. In short, all human bodies and minds and behavior will be in the Internet. So much of our world will be virtual via smart devices, even robot clusters, performing human functions. The new, required standard architecture will be a Panopticon—a circular prison with the guard tower, human or digital, in the middle able to observe every move of a prisoner, patient, student or hotel guest. The Digital key will be needed for access. Indeed, more than a password, the Digital ID will control the material world. In a real sense, it's already here when Alexa does what you want. "She" knows so much about you already!"

"Ok. So, what are you really saying, Chin Chin?"

"In effect, a central leadership can control the discussion. Period. It can purge critical thinking. It defines free speech and ignores the Constitution. It controls everybody's movements. In China, this is what the SCS is all about. For no reason at all, the SCS Controllers can send a person to an Intellectual Lithium Mine, never to be seen or heard again. So, be prepared. It's a world where words like "diversity," "equity" and "inclusion" are verbal traps set by the Socialists and ultimately the Communists."

Andrew nodded his concurrence because his partner was on a roll and had the truth on her side.

"So, Chin Chin, having said all that, is VSI on the right track by trying to get preteens to stop using these smart devices (the doorbells to Googleland) in order to wake up to the realities of the world they will inherit?"

"Yes, Matt. And the task gets larger everyday as more and more youth are addicted to that doorbell."

"This ends our discussion. Please join me in the bar. I need a stiff drink," requested Matt.

On the Gulfstream the next morning, Matt was still shaking his head about the potential new world of Digital ID to destroy individual freedom in America. And he knew that if the Central Bank (Federal Reserve) enforced cryptocurrency, control of the electorate would absolute. He now had enough information to accelerate VSI's initiatives with funding from the Board of Directors back in Miami.

10

MIAMI III

Three days after the Gulfstream landed in Miami, right on schedule, the following friends began arriving at the Key Biscayne compound for the VSI Board of Directors meeting:

- Evelyn Ebbert, (Ebbie) lawyer and Secretary to the Board
- Alexander Murray, President, VSI
- Harvey Rubinstein, CPA and Financial Advisor
- Dawn Langford, President and CEO of Padlockers
- Cole Langford, Vice President of Padlockers
- Yousif Latif, Prince of Dubai; Investor with the Flynns
- Bill Cavanaugh, Vice President, Sales and Marketing
- Maurice Cavett, Sr. VP, Alcatel S.A. of France
- Anthony Bracca, VSI Production Director, New Braunfels
- Murray Flynn, Deputy Chairman of the Board

Promptly at 10:30 the members took their seats around the dining room table for the secretary to call the roll. Alex welcomed everyone.

"Welcome, my friends, we've got a lot on the agenda with exciting news. Our plan is hold this meeting until noon. We'll have lunch served by the pool until 2:30. At 3:00PM, a Maybach will be available to take anyone to South Beach to see what a diverse culture looks like. Let me tell you, you'll come away shaking your head and wondering where mankind (anykind!) is headed. Happy Hour will be in the library and dinner here at 7:00. All spouses and partners are truly welcomed. Now, without further ado, Chairman Flynn will provide the status, complete with our recent trip to Singapore."

"Thank you, Alex, I look forward to a very productive meeting. I will wait to make a couple of special announcements until Happy Hour when we can lift a glass of champagne to celebrate. As Alex said, we just returned from Singapore. We had a briefing by our close friend Chin Chin Po and her partner, Andrew. In a secure setting, the couple was quite frank about current events in China. The more they talked, the more I felt an urgency in our mission to curb the deleterious effects of unchecked social media. Over the Pacific, I wrote down some of the key points they made. Here's the short list:

- Over 90% of all S-M devices sold in this country are made in China.
- The CCP uses S-M to silence ANY opposition

- Censorship is embedded in all S-M services. (Tens of millions of cameras report every move.)
- A Social Credit System (SCS) controls the actions of hundreds of millions of people in the big cities, and it's spreading to the small ones as well. Harsh penalties are assessed for stepping out of line. And the penalties are never forgiven so the individual and family are blackballed for eternity!

(Everyone in the room shook their head in disbelief.)

- Social media is used to infect over 50,000 Chinese students at American colleges and universities with totalitarian political views to fan racism as a smokescreen to steal our technology.
- All S-M devices can Dumb Down our youth into accepting a central government's leadership for life. (Sounds like Stalin, doesn't it!!)
- S-M is used to justify DEBT TRAPS in countries like Mexico and Canada to build the communications infrastructure to be able to control the citizens. In part, the country of Mexico repays China by allowing illegal drugs to be made in the country for illegal importation into the United States. And, of course, Fentanyl here is used to further Dumb Down Americans.
- China owns Hollywood and controls the messages of streaming videos

Let me be quite blunt. An SCS, augmented by a Digital ID and Cryptocurrency, dooms our republic.

We've made presentations in Washington to tax and tariff the Chinese products and their services. We should learn about our efforts any day now.

We can apply our incredible weapon, SHADOW, against the social media production in China in order to allow, indeed, force our Flip Phone production to be a commercial success."

(Nodding heads surrounded the table.)

"Matt, thank you for that scary future forecast! Actually, your remarks are a perfect segway to our next topic: Status of the new factory. Tony, what's new in New Braunfels?

"Thank you, Alex," said the director as he walked to end of the table, he motioned to Ebbie to project his slide show on a screen. He ran through dozens of slides showing the progress from the ground up. He highlighted the automated assembly line and complemented the teams of Japanese engineers. "And I'm particularly proud of our workers who have the highest Work Ethic on the planet." So, my friends, please plan to join me next month for our ribbon-cutting ceremony. By then, all 126 employees will be trained, certified and able to produce the Alcatel Flip Phone. Our quality assurance staff has been augmented by technicians from France. Merci!! Merci!!

Preliminary production runs show an industry-leading low rejection rate of 1 per 1250. So, with confidence, we can accurately predict final construction costs at the docks, ready to accept the Amazon trucks. Thank you for your confidence in our entire team to make this dream a reality. Any questions?"

(There were none; only applause and Bravo's.)

After a coffee break, Murray went to the end of the table to report his and Maggie's briefing at Georgetown University. He summarized his half-dozen scenarios using smart phones and the hostile reception by the audience with "every soul addicted to their smart phone."

"The Bottom Line is exactly what my father has been saying for the last year. American needs a revolution against these smart phones which begins at home and in grade school in order to wean the next generation of these hand-held Chinese weapons. The Flip Phone can be a spark which ignites the flame for change.

Alex then introduced Dawn to report on the joint project with Amazon and the secret project to develop SHADOW.

"Thank you, Alex. As you know, we have dedicated the last 4-months to VSI projects. We're believers, and we feel the need for an advanced voice-activated Flip Phone. I'm glad to report progress on both fronts. First, in the middle of the table is a speaker. Bill, would you please flip the switch on your side."

"Hello, Dawn, glad to be with you today. How can I help?

The board members smiled with delight!

"FLIPA, please report on the status of Valerie's homework."

"She has completed her math assignment, and all of the answers are correct."

"Thank you."

"FLIPA, Rodney had a paragraph to write about his recent field trip to a space museum. What is the status?"

"Rodney has completed his draft, but it has two punctuation errors, both are run-on sentences. I alerted him, and he is making the corrections."

The Board members simply shook their heads in delight.

Dawn continued, "Friends this demonstrates the power of AI-based voice synthesis. Our goal is to have a commercial version of the software ready to load on the phones in Texas within 60-days. We think it will take the market by storm!"

"Dawn, the weather in Texas is sunny. There is no storm."

"Thank you, FLIPA. Goodbye."

"Goodbye, Dawn."

"We have not applied for patents because we do not want the Chinese to steal our technology."

"Dawn, that was EXCELLENT. It's clear that the combination of the Padlockers and Amazon is an incredible social change agent. And progress on SHADOW, my dear?"

Dawn asked Ebbie to start a video. The two-minute video showed a laptop being examined by a gray-haired repair technician. "I don't know why it won't work. I've tried every diagnostic routine we have with no results. Even our super-smart software doesn't know what's wrong. As it is, it's worthless and a true mystery to our staff."

Dawn nodded to Ebbie to stop the video as she said, "It's no mystery to the Padlockers. We've downloaded an invisible "E-Worm" which makes the device incapable of accepting electrical power. The laptop (or smart phone) is dead until we remove our E-Worm. And let's be quite clear: our E-Worm can replicate at the speed of electrons shutting down the very power grid servicing the laptop's location! Any questions?"

There were none, just stunned silence.

"Dawn, what your team has developed is the most powerful weapon in human history," said Prince Latif. "If I understand you correctly, a nation's power supply could be shut down until certain demands are met. With respect to VSI, Chinese power grids could be shut down until they stop making smart phones for American companies like Apple or Google. If this were to happen, the advanced VSI Flip Phone would be the logical answer for mobile connectivity in America."

"Prince, you're absolutely correct. And because the national grid remains dead due to our NetBomb, the source of the E-Worm is undetectable. Period. If we ever employed it, we could plead innocence, free from attribution. Only a few of the senior staff at the Pentagon, and a handful of cyber gurus know it exists. Foreign missiles can't fly without computers, dozens of them. It's the ultimate security tool in the nuclear age. But as Matt has formulated: a small demonstration in Beijing could be enough for the Chinese to stop exporting brand name smart phones to America.

Our factory in New Braunfels would be the model for the next generation, perhaps generations."

The tall, stately woman walked over to Ebbie's laptop, extracted the thumb drive and sat down. The members clapped as Matt introduced a Motion with five parts:

- "Complete the contract discussion with Alcatel
- Authorize the factory ribbon-cutting ceremony
- Continue the FLIPA voice systems development
- Continue the trials in primary schools in Virginia
- Keep SHADOW a national secret only to be used with national level approval based upon Intelligence Community (IC) understanding that the goal is to return device manufacturing to the USA."

The motion was seconded and approved without additional discussion.

The afternoon saw the members and their spouses or partners playing water polo in the swimming pool. The trip to South Beach was cancelled.

11

LEESBURG II

Two weeks after the Board meeting, President Murray received a call from the President of the Lee County School Board.

"Mr. Murray, today ends our 30-day trial at the Lincoln Elementary School. Would you be able to come the school during the lunch hour today to talk with principal and the home room teacher to talk about the results?"

"Yes, I can be there. Will you be able to join us, Mrs. Brock?"

"Yes, I've cleared my calendar. I will be there. As you know, I'm keenly interested in the results. My twin teenage girls are addicted to smart phones, and I'm very concerned."

"Yes, I relate personally. My son was in La-La Land on his device for four-months until he found a job. See you at noon."

At 11:45 a VSI T-Pod landed on the baseball diamond behind the school. Alex powered down the craft as Maggie Flynn gathered her computer carrier. The duo was met by

the principal at the security checkpoint. The trio paused for a moment to inspect the Flip Phone cubbie. There was still one phone attached. The principal said: "That's Mr. Robert's, our chief custodian's phone. He wanted to be part of the trial."

At that moment, Mrs. Brock cleared security and joined them. With an air of excitement, they walked to the cafeteria and found a quiet corner table. A minute later, a slender, tall middle-aged woman with long silver hair walked up to the table and introduced herself as Miss Lambert, a 7th grade teacher with Homeroom responsibilities. Proper introductions were made as three coffees were sugared. President Murray opened the meeting.

"Thank you for your consent and assistance in this trial. Our recent research convinces us that the smart devices can actually consume a person such that it becomes a necessary appendage. VSI is really concerned about the impact on the nation, one student at a time."

"Mr. Murray," began Miss Lambert, "we followed the trial's instructions precisely as provided. My class underwent the training. Our daily records confirm the locations and use of the prototype VSI Flip phones. I've summarized our experience on this sheet of paper:

- At first, three (3) students were very nervous but calmed down when the purpose was repeated
- After a week, more participated in classroom discussions. They actually raised their hands to answer questions.
- Students looked up from their desks and actually made eye contact with me and other students

- Complete sentences were used; not just the usual text spurts"
- Midway through the trial, two young boys came to me to ask to be excused from the trial because they were being bullied by siblings at home. When I explained that they were the lucky ones who were no longer slaves to the device and could begin to live life, they agreed to continue in the trial.
- One student made an emergency call when frightened while walking home from the bus stop.
- One student, the biggest bully in the class, actually came up to me and told me he could live with the device. "I have computers at home and can use them when I need to do research. I don't feel an urge to access the Internet when carrying the Flip Phone."

In short, Mr. Murray and Mrs. Brock, I see a change in the classroom after only one month. I'm starting to be an advocate of this new, smart approach to mobile devices."

Mrs. Brock asked if there were any questions for Miss. Lambert. There were none.

"Miss Lambert," said Alex, "your participation and summary are appreciated. This is precisely the feedback our company needs to be totally responsive to the you and your students. Thank you very much."

At 1:30, the T-Pod was airborne at 150 feet headed east back to Alexandria. As the craft flew south to avoid the Dulles Airport airspace, Alex and Maggie gave each other the "high-five" salute. With luck, this would become a standard program in that school district and perhaps beyond.

12

PENTAGON

On a cold November morning, a black Suburban picked Matt, Alex, Cole and Dawn at the VSI offices in Alexandria for the 5-minute trip to the Pentagon. The foursome was ushered through security and escorted to the 5th Wing and into the Assistant Secretary of Defense's conference room. The walls were lined by senior military and staff personnel, cybertechnology experts, a few of which were recognized by the Padlockers. The 3-star general entered the room and took his seat at the end of the table with Matt and Alex on one side and Cole and Dawn on the other side.

"Welcome, my friends. This is quite a moment for me. I've been involved in cybersecurity for over twenty years. My doctorate is in the specific domain where you, Cole and Dawn, created the SHADOW program. Everyone in this room confirms that this software is incredible. In complete secrecy, and I mean COMPLETE secrecy, this code could have the largest impact of any on Earth. This country is extremely proud of the team of developers and VSI for initiating the program to make this day possible.

Let me quite frank. If we were to announce it, you would hold the Nobel Prize for Peace. Congratulations!"

The general shook the hands of the four guests. All of people in the room rose to their feet and applauded.

"And let me also say that this code (He reached in his pocket and produced a thumb drive.) is an atom bomb in the wrong hands. SHADOW will be classified at a level higher than most of the people in this room. Only my boss will be told of its existence. Only the President can announce its existence, much less use it. I hope that it is absolutely clear to everyone in this room…a room with no smart phones or recording devices of any kind."

"Yes, Sir!!" was the total response.

"So, you can appreciate my concern when our chief guru, Dr. Sebastian, informed me that you want to use it to kick-start the national production of Flip Phones! Matt, I ask you: "What do you have in mind and why?"

"General Taylor, VSI exists to rescue to the next generation from the addiction of mobile social media. You may remember a few months ago we spoke about this addiction to the smart phone and our 4-step plan to end the problem over the next decade."

"Yes, of course. Your words did not fall on deaf ears!"

"Well, Sir, we would like to perform a SHADOW surgical strike on the Apple IFone plant in Zhengzhou, China west of Beijing."

"Oh?" with raised eyebrows around the room.

"Yes, the Foxcomm factory produces over a million of these very smart phones annually. Some are sold in China, but the bulk of them are exported to our shores. We think a surgical production stoppage would force Apple executives to seriously consider and ultimately accept a production grant for a new facility here along with a guaranteed purchase by the Federal Government of, you guessed it, Flip Phones. We, of course, would bid on the contracts as well.

"Mr. Flynn, on the surface your plan seems idiotic. But upon further reflection, it has merit for a reason you may not have thought about."

"General, you've really got my attention." Said Matt as he made a quick ceremonial hand salute.

"The military doesn't advertise the fact that we have a huge problem with smart phones. All branches of service have to constantly deal with social media in the ranks… in the barracks, in formation, in training, and, most recently, in combat. This is particularly the problem with new, young recruits at the Marine Corps depot at Parris Island in South Carolina. These youngsters get off the bus with their heads down looking at a device. It's the first thing the Drill Sargent confiscates. We've even installed cubbies at O-clubs to experiment with person-to-person communications."

"Thank you, sir, for that enlightenment. I/we had no idea."

"So, if, and it's a BIG if, you can prove that SHADOW can do its job without any retribution whatsoever toward this country, we might be able to craft a credible surgical strike. Of course, if somebody, anybody, points the finger at us, we'll direct them to Bermuda."

"Yes, Sir. It's understood."

The VSI trio at the table nodded their concurrence of the requirement.

"My input to this plan would be funds from classified accounts."

General Taylor smiled at Dawn: "Like you, some people in this room know exactly what and where I mean!"

We would put out a Request for Proposals (RFP) for a huge number of Flip Phones made in this country. We might even offer a factory site on a closed base. Our Corps of Engineers could plan the site. We could handle the personnel recruitment requirements. This "pump-priming" just might be attractive to a company like Apple with the thought of a virtually guaranteed future market."

Matt thought to himself: "My God, there just might be a pony found in them thar hills!"

Alex looked at Matt, got the nod of silence, and said, "General, VSI is ready to support this endeavor in anyway we can. We have the desire to serve because no action now will end badly for our country."

"If your people can convince us that a clandestine SHADOW has a high chance of success, I'll approach my budget folks. No word goes out of this room. My boss will make the ultimate decision. I will inform you of our decision within ten days."

With that, General Taylor rose, announced that the meeting was over, shook his guest's hands, turned around and walked into his office. As Cole was leaving, he overheard the General to tell an aide, "I'm sure glad I can retire soon."

13

PENTAGON II

After nine days of rigorous testing of SHADOW in the Pentagon's CyberSim laboratory, the Padlockers were told: "Ok, this is your final test prior to a decision."

An hour later, it passed, and the couple did a "high-five" salute.

Later that night they were on a Hypersonica flight to Singapore.

HYPERSONICA

JASON O'NEIL

After a few hours sleep at the Raffles Hotel, they met Chin Chin and Andrew at the American Consulate. In a secure basement conference room certified "clear" by

CIA employees, the Padlockers were briefed by the couple from Hong Kong who could not be seen by the Chinese leadership. They came with yesterday's pictures taken by friends in key places.

"Yes, that's the factory configuration on which we did our simulations," said Cole. It was clear to the foursome that the software would perform as tested.

During a working lunch, a secure videoconference was set up with the Pentagon. Much to the Padlocker's delight, General Taylor and the entire Joints Chiefs of Staff were present. Cole and Dawn flipped a coin to see who would lead the presentation. Dawn won the honor.

"Distinguished Officers, we believe we're ready for the subject mission. We have executed the software on a true replica of the subject location. It stopped immediately and was unable to be restarted. When key printed circuit boards were replaced by the Apple technicians, the silent worm continued to make the machines inoperative. Work around solutions attempted by your teams there and here have all failed. And the source of the stoppage could not be traced, even by real-time supercomputers. Gentlemen and Lady, in my professional opinion, I believe our product is ready for "Show time."

General Taylor then had the Zoom camera focus on one of the men sitting against the side wall.

"George, are you OK with the plan and timing?"

The stout figure with a bushy moustache and thick spectacles simply nodded his head up and down. The man was never introduced to the people in the Consulate's basement. But the man's motion was enough for the General Taylor to sign off and the video went dark.

The four VSI employee/consultants looked at each other as Cole gave the "Thumbs-up" sign.

An hour later, the foursome enjoyed Happy Hour in the hotel's world-famous Mahogany Lounge. It was followed by dinner and small talk prior to retiring for the evening. Dawn slept with two thumb drives in her cleavage. The next morning the Chinese-American couple returned to the former British colony. The Padlockers boarded a Hypersonica flight to Dubai. A secure text on a new "special" Flip Phone informed them to activate the SHADOW signal when they receive a coded command.

14

DUBAI

As Cole and Dawn walked out of the Hypersonica jetway at the Dubai International Airport, they were met by Prince Latif. The trio was driven to the Prince's compound across the street from the main city mosque. As they walked into the foyer of the mansion, the prince said, "My home is your home. Please be comfortable."

Dawn looked up at the two-story foyer completely covered in gold. She said to her husband, "This is going to be good!"

After freshening up, the couple followed the prince's orders and met him in his library. The trio sipped champagne as they received a call from Matt wishing them success in their mission. He ended with a promise to return to Key Biscayne within ten days for the next Board meeting and the urging to "be sure to take advantage of your host's hospitality!"

In the hour before dinner, the prince explained the geography of the Emirate in detail and the steps to be taken should they get "the text" in the next two days. Prince Latif assured his guests that whenever the mission was activated, the weather would be sunny for the 33-mile T-Pod ride to the east coast at Limah. It was the site of a multi-megawatt transmission tower with line of sight to a DoD satellite in space over the Orient.

Dinner was a quiet, reflective occasion as the prince lamented how social media had become "a disease in the Emirate." With after dinner drinks in hand, the trio walked back into the library. The prince put his drink on a table and walked over to a picture of a sultan on the wall, swung it open to expose a wall safe and put his beckoning hand out to Dawn. She smiled and knew what he meant. She took the thumb drives from her cleavage and handed them to him. The Padlockers watched as he placed it in the safe, closed the door and spun the tumblers. As the picture was swung back into place, he said, "The safe is being lowered into a huge bank vault in the sub-basement for additional security. Now we'll all sleep better tonight."

The prince was right. A good night's sleep came easily. They weren't even disturbed a "ping."

After breakfast, the couple made calls to Bermuda to assure that everything was going as planned. Mid-morning splash time in the swimming pool was followed by some late morning delight in their bungalow. For lunch, they met the prince in his conservatory. They had a long call with Matt and Alex. The prince promised to be in Texas for the ribbon-cutting next week. As lunch was finishing, the prince gave each of the Padlockers a small envelop. Dawn opened hers first. It was a credit card for the Gold Souk bazaar only two miles from the compound. Cole opened his. "Incredible! What a wonderful surprise!"

A half-hour later, a Maybach pulled up to the mansion to take the couple to the famous Souk. "Enjoy, my friend. The card is unlimited…a thank you from the management of VSI and me. See you for dinner."

At 6:00 PM, the trio deplaned from a gold T-Pod at the base of the Burj Kalifa, the world's tallest building. On the 118th floor, the prince introduced his friends to national dignitaries as the trio made their way to the prince's table. Dawn was truly sparkling in a black V-neck silk pants suit with a heavy gold necklace shining on her bodice. Cole marveled how beautiful this woman..HIS WIFE…really was. The seven-course dinner was a private banquet.

The T-Pod ride over the sparkling desert ended on the lawn of the compound. The prince powered down his gilded craft and led the way to a gazebo for after dinner drinks of sixty- year-old Cognac. Dawn commented on how she got warmed with each sip. Prince Latif quipped: "That's right. I even have it in my ski poles when I'm in Switzerland. At 11:15 under a full moon, the couple walked to their bungalow. Sleep came easily for the nude couple between gray silk sheets.

At 6:30 the next morning, Cole wrapped his hands around a silver coffee mug and smiled as his love wiped the sleep from her eyes. He felt a buzz on his Flip Phone

in his robe pocket. He pulled out his phone and saw there
was a text message. He scrolled down and clicked "OK" to
read "Show Time."

Dawn sprang out of bed, wrapped a towel around her
slender body and took a sip of Cole's coffee. Ten minutes
later, the couple walked across the compound past the
swimming pool and into the conservatory. As they walked
past the lush green plants, they could see the prince opening
the safe in the library.

"We have food in the T-Pod; let's go!"

The noble showed his pilot skills as the craft traversed
mountain tops and narrow valleys headed east to the
transmission tower. The T-Pod landed on the parking lot
next to the automated, unattended control shed. Using a
remote control, the prince unlocked the door to the shed.

The trio entered and immediately went to the control
console. Cole unpacked his laptop as the prince powered up
the transmitter. (Later, he would confide that his company
built it so he knew everything necessary to power up.) Cole
looked at his Breguet watch to see 8:55AM. He beckoned
for Dawn to give him the small piece of paper Chin Chin
had provided in Singapore. From his smart phone, he
entered the Foxcomm plant's exact geographic coordinates
into the laptop as well as the elevation, azimuth of the
satellite and the requisite transmission power. He nodded
to Dawn who then inserted the thumb drive into the side
of the laptop. He pointed to the Enter Key. With a strong,
focused demeanor, she pushed the key. The transmitter
made a low whirling noise as it powered up. The laptop
display showed an increasing signal strength and symbol

of the transmission energy wave. In only the two-minutes, the spacecraft's focused, high-power beam delivered the software precisely to the Foxcomm facility, and the mission was completed. Cole then took the thumb drive out and handed it to Dawn. The prince took out a container of sanitary wipes to wipe off any equipment the trio had touched. The station powered down automatically as the prince predicted.

"We're done. Let's go."

The trio left the shed, wiped the door latch and walked to the T-Pod. The prince then pointed to Cole and said: "Your turn to pilot, my friend." A broad smile come over his face as he accepted the key. Cole proved to be an excellent pilot, much to Dawn's relief. As the craft entered Dubai city, the prince assumed control and adroitly landed in his compound. With the craft powered down and secured, the trio walked to the library to put the thumb drive in the safe. As he swung the portrait back into place, the prince said, "Congratulations, you've just started a new chapter in the history of mobile S-M in America!"

The look on the Padlocker's faces was priceless.

15

CUPERTINO

In the headquarters of Apple Computer is the global production command center with a video wall showing all daily deliveries and any production outages. A light rain was falling in the Bay Area, as a technician settled into her chair for the night shift. Her specialty was supply chain visibility and logistics support for the largest social media device seller on the planet. She logged in to start her shift and on a particular Bill of Lading for 10,000 I-Phones coming out of central China destined for Los Angeles. As she reviewed her status screen, she noticed that a location on the map was flashing red. She accessed several cameras at the Foxcomm factory in Zhengzhou, but nothing happened. The screen was blank. She powered down and restarted her computer. She again accessed the site, but there was no signal. Per procedure, the technician picked up the phone and dialed the contact number. There was no answer. She then called the plant director on his personal cell phone. This was the last resort which had never been required since the plant opened five years ago.

"I know why you're calling. The plant is down, completely dead. We're working the problem. Everything around us is lit up and working fine, but we're dead. So glad we have panic bars on the exit doors so all of the employees could exit the buildings. We have calls into the power company. They were not aware of our problem. Please inform management that we're working the problem as best we can. Trust me. Our entire staff is brainstorming and trying to figure out what has happened to the plant and even our back-up generators. It's as though a virus has attached every system and can't be isolated for removal."

It was an "OH, SHIT moment for the young lady who immediately called her boss.

"Nancy, there's a problem in Zhengzhou. The plant is down, and we're counting on these deliveries of the latest phones for Christmas. I've spoken to the director who assures me that they are doing all they can to restart production."

"OK. Janice please text me when it has restarted."

The call was never made.

Six days later the Apple Board of Directors met for their regular monthly meeting at the headquarters. The members quizzed the President. "What do you mean the plant's still down. Their management is brilliant; they were all trained at UCLA. This is unacceptable!"

"Yes, gentlemen. It is unacceptable, but our hourly calls to the leaders at an off-site location have not given us hope of a near-term solution. It's almost like a virus has infected every cell of an organism. Nothing turns on. People are walking around with the lights on their cellphones. One member of the board shook his head and wondered if the company should re-start production in America.

"What do you mean?"

"Well, as you remember, we moved that production from Mumbai to Zhengzhou because that region of India was levelling more and more restrictions on us and our employees. It was some social credit system, and it stole much of our profit.

Yesterday, we had a visit from the U.S. Department of Commerce and the International Trade Organization. They alerted us to the following developments:

- A Bill in Congress would slap a 26% tariff on all digital devices imported from Asia, meaning China
- Federal tax would be increased at least 22% on all companies importing "smart devices of any kind" from Asia
- No cost, free interest grants being made available to a company which relocates out of Asia to America. The grants cover the entire production development process
- New production facilities could be developed on unused military bases with a 99-year lease with the free assistance of the Army Corps of Engineers

Several of our business partners suggested that we look into the program especially since California has voted to leave the United States. It means a huge uncertainty, especially when Sacramento is broke. And when this happens a border wall will be constructed and new tariffs imposed on our products. Two companies I know are in Arizona today inspecting potential sites."

"Wait a minute, Ira, there must be a catch. This sounds too good to be true. The federal government is never this rational. What's the logic behind this movement?"

"Well, I've only seen the preliminary documents, but the intent is quite clear. National polls have proven that these mobile smart devices are dumbing our youth to become non-productive wards of the state dependent upon new Socialist handouts to maintain control over the citizenry."

"But politics be damned. How do we increase our profit?

"The Army will soon issue an RFP for cell phone, millions of them."

"Great. We can bid on it, and our friends in Washington will see that we get the contract."

"No, Sir. You're wrong. The contract is for Flip Phones made in America."

"What?"

"Yes. The military's recruits are addicted to our smart phones. The new soldiers can't survive without them. It's a huge discipline problem. The RFP will highlight a training incident when a Marine lost his life texting on a smart phone. It was probably one of ours."

"Ira, this is a real challenge."

"Yes, it is. But we can make the most of it. As I see it, we can have a four-step program which can really benefit our company while helping the country…some really good public relations here. We can:

- Relocate the robots and control equipment from China to Arizona, Texas or somewhere in the South
- Create a state-of-the-art manufacturing facility at the taxpayer's expense
- Split the production into two lines: one for the Flip Phone and one for our new devices for export
- Bid on the military RFPs to have large, guaranteed sales for a long time to come

And, all this avoids onerous taxes and tariffs as well as the uncertainty of production in China or even the new California called "Calicopia."

At the very moment the board members were mulling over the suggestion, an aide slipped a note to the chairman.

"Gentlemen, I just received word that the Chinese have declared the Foxcomm plant area a "Dead Zone" where all the equipment must be removed. Unless Ira say's otherwise, this could be a huge tax write-off creating funds to relocate here at no expense."

"Sir, I believe we could work that. Yes."

"It's absolutely clear that this is an opportunity which we must immediately research, and, if warranted, bring it up for a vote, even in emergency session at our off-site meeting next month."

16

KEY BISCAYNE

Two months after newspapers reported the mysterious outage of an Apple I-Phone production factory in China, VSI had its monthly Board Meeting at Key Biscayne. Fresh from the ribbon-cutting ceremony in New Braunfels, Matt, Alex and Prince Latif gladly handed members the first Flip Phones off of the assembly line. The plant's director, Tony Bracca, was asked to stand and be recognized. Matt went over to him, shook his hand and gave him an envelope. Everyone cheered the occasion. A special guest, the manager of the Amazon Fulfillment Center in Schertz was also asked to stand. As Matt was shaking his hand, the manager announced, "I'm proud to report we can now handle 50,000 a month and more after our expansion is completed in 90-days."

(Bravo, Bravo)

Matt returned to the head of the table and began his remarks, "My friends, I have trouble finding the right words to express my appreciation for your efforts to make the VSI dream become a reality. Over the next two days, I will highlight each of your accomplishments with pride.

But let me kick-off this meeting with a couple very special words of gratitude.

First, will the distinguished couple from Bermuda please rise. My friends, our company is impacting world history one hand-held at a time thanks in large measure due to the Padlockers."

(A roar of approval could be heard throughout the compound.)

"They were instrumental in shaping corporate decisions to produce Flip Phones in America. Prince Latif was instrumental in the execution of the mission based upon the trusted word from Chin Chin and Andrew. The nation will probably never know just how important your efforts were for the globe! Let's be honest. The Padlocker's invention is so important and so powerful, that the thumb drive is guarded at Fort Knox. All other versions of the software have been destroyed."

Matt looked at Cole and asked, "Right?"

"Yes, sir. We don't even have a copy of the source code in Bermuda."

"Perfect."

"And, I would be remiss if I didn't recognize the courage of our newest Board member, recently retired General Taylor. Sir, we salute you!"

(Genuine applause and "Bravo" filled the room.)

"Thank you, Matt. You're most kind, and I have two items to report: First, Alcatel and Apple are the winners of the recent RFP according to my sources."

(All eyes were on Vice Chairman Cavett as he stood and accepted the applause.)

"And, secondly, the Marine Corps Commandant just sent me a kind note telling of a new focused attitude of the recruits at Parris Island who have taken part in a Flip Phone trial."

"Guys, we're on a roll. Let's keep going around the table to discover more good news impacting our company. Alex, as I remember, you have good news on two fronts."

"Yes, Matt, I do. First, three other smart device makers, Google, Nokia and Phillips, are reducing their production in China and looking for factory sites here as Apple just announced a new campus on the Army base, Ft. Rucker, in Alabama. We love this competition. Please look at this new commercial for FLIPA"

Ebbie ran the one-minute video much to the delight of the Board. It showed how FLIPA helped a young student complete her homework.

"And soon we'll announce a new model of our phone which has an extra battery to run a taser by voice command for personal protection."

"And, my friends, I have another good news headline. The government just reported that the money from the tariffs on Calicopia will actually pay for the 600-mile -long border wall between our Republic and the new nation."

During a bio-break, the prince motioned for Matt and the Padlockers to join him in the conservatory behind a bubbling fountain.

"Guys, I want to tell you that last week we razed the communications tower at Limah in favor of a bigger one further down the coast. With this action, our mission should remain a "cold case" for a long time! And, if I'm successful in a meeting with Royalty next week, our Flip Phone will be required for every student in the country. They will get it free and have to use it, or the family will pay penalties."

(The trio smiled and appreciated how powerful this man is his country.)

A few minutes later, Matt reconvened the meeting.

"Friends, I'm happy to report that China has recently lowered it's smart phone prices for the Third World markets."

Chin Chin confirmed the report and added: "And our sources confirm that Beijing is running secret trials of Flip Phones in Mongolia. Tinshin Wei has confirmed the trials and something even more important: riots in remote cities against the social credit system. You won't find this in the newspapers, but it's big, very big, news. A limited-function Flip Phone would greatly reduce the social terrorism conducted by Beijing. Please stay tuned for major events in the coming months."

"Yes, we will anxiously await more news like this out of the Asian dragon. Now, in many ways, I've saved the best for last."

He nodded to Ebbie to display a graphic of a newly launched children's book.

"Look at the author's name,"

(A roar of appreciation was heard as the group realized the author was Heather!)

She rose to the acclaim and was asked to describe it. She pointed to the graphic and said, "Flipphonia" is a parent's guide for the use of Flip Phones."

To the delight of the audience, she described how the Flip Phone and FLIPA could be a new educational tool. The book is filled with quotes from parents about the positive impact upon children at home and in the classroom. It's like a re-birth of the Work Ethic, almost extinct in some parts of the country."

She handed out copies of her book around the table and announced with some humility, "It's 4.5 stars on both Amazon and Barnes & Noble and selling very well." Hugs greeted her as she made her way back to her seat.

"See, I was right. I saved the best for last! The board will meet for an hour this afternoon to vote on a couple financial motions. In the meantime, our home is your home; please enjoy. Everyone is welcome for Happy Hour by the pool at 5:00 PM."

As the members were filing out of the dining room, Harvey, the financial guru, pulled Matt aside to alert him to a motion to award some of the year's profit to inner city school districts to kick-start new FLIPA WAVES. He simply smiled and said in a low voice: "Thank you, Lord."

17

BERMUDA

Because Cole and Dawn are the only two people on Earth who can reproduce their history-making software, the U.S. Government requires that they take separate conveyances whenever they travel, just like the founders of Coca-Cola. Consequently, Cole flew from Miami to Bermuda to get ready for the Board of Directors and guests arriving aboard Prince Latif's 288-foot megayacht, Wanderlust.

At 12-knots cruising speed, the huge blue ship takes three days to sail the 1,000 miles from Miami to Hamilton, Bermuda. Aboard the floating palace were:

- Matt and Heather
- Murray and Maggie
- Chin Chin Po and Andrew Ho
- Dawn Langford
- Maurice Cavett and Jacqueline
- Alex Murray and his wife
- General Taylor and his wife
- Harvey Rubinstein and Heidi
- Ebbie and her husband, Jason

The prince has a crew of twelve plus three young female escorts. During the trip, the guests were treated like royalty with exquisite dining, sports, spa treatments and even skeet shooting off the fantail. The well-known vessel docked in Hamilton late on the third day.

On the north end of Bermuda's main island is a small island, formerly called Smith's Island, but now Savoir. It was purchased by the Padlockers as a dedicated spa colony to attract and keep the best super cyber scientists on the planet. It worked. Thirty software engineers and their families relish the lifestyle while creating unique software deliverables. The two-mile long island has condominiums on the coast around a central Green and small retail arcade. It has its own clinic, supermarket, sports venues and power station with back-up.

Wanderlust's guests were brought to Savoir by high-speed tenders. Cole welcomed them with room assignments and an itinerary for the weekend. Three T-Pods were on the Green available for guests to visit anywhere on the island paradise. The only firm requirement was to dress for formal dinners, one at the famous Rosedon Hotel and the other a banquet aboard the ship. Most of the guests swam in the warm water, played croquet on the Green, rode scooters around the island and shopped for high-end wristwatches in the duty-free zone in Hamilton.

On the first evening, the Governor himself joined the party at the Rosedon. The event was truly joyous. The circular buffet table had large Flip Phone ice sculpture in the middle. And the Governor was glad to receive a free Flip Phone for every school child under the age of 19 and teacher on the island. Training seminars would be set up for each of the schools.

The second day saw most of the guests play golf at one of nine golf courses. The foursomes returned to Savoir in time to freshen-up for the banquet aboard the Wanderlust. The prince greeted each couple as they boarded his floating palace. The six-course dinner was as good as any in France according to the French couple.

During dessert, Matt rose to give a short dinner speech. He thanked everyone who made the occasion possible. As a special surprise, he offered a condominium apartment to each couple in the same building as the one customized for the Padlockers. The audience was thrilled. A drawing was held where Heather drew a name out of a top hat. The winner, Andrew Ho, was awarded a new Bentley Continental 24 automobile. The car would be shipped to Hong Kong. The couples danced the night away. During the occasion, each couple expressed their appreciation to the Flynns and Prince Latif.

The next morning came all too soon for those who caught commercial flights back to the USA. The Padlockers watched seven planes disappear over the horizon as they sipped Baldie rum cocktails. At their feet was a fuzzy gray puppy given as a gift from the Governor's office. The couple looked at the ball of fur and asked each other, "What shall we name him?" Their joyous response was immediate and simultaneous, "SHADOW."

www.ingramcontent.com/pod-product-compliance
Lightning Source LLC
Chambersburg PA
CBHW030939240526
45463CB00015B/582